DEEP BLUE HOME

Books by Julia Whitty

*A Tortoise for the Queen
of Tonga*

The Fragile Edge

Deep Blue Home

DEEP BLUE HOME

AN INTIMATE ECOLOGY OF OUR WILD OCEAN

Julia Whitty

MARINER BOOKS
HOUGHTON MIFFLIN HARCOURT
BOSTON   NEW YORK

First Mariner Books edition 2011

Copyright © 2010 by Julia Whitty

www.hmhbooks.com

Library of Congress Cataloging-in-Publication Data
Whitty, Julia.
Deep blue home : an intimate ecology of our wild ocean / Julia Whitty.
p. cm.
Includes bibliographical references.
ISBN 978-0-618-11981-3
ISBN 978-0-547-52033-9 (pbk.)
1. Ocean currents. 2. Deep-sea ecology. 3. Whitty, Julia—Travel. I. Title.
GC231.2.W49 2010
551.46—dc22   2009047558

Book design by Brian Moore

Printed in the United States of America

DOC 10 9 8 7 6 5 4 3 2 1

Lines from "Lament for the Dorsets" by Al Purdy, from *Beyond Remembering:
The Collected Poems of Al Purdy.* Copyright © 2000 by Al Purdy. Reprinted by
permission of Harbour Publishing.

Lines from "No More Fish, No Fishermen" by Shelley Posen. Copyright © 1996 by
Shelley Posen, Well Done Music, BMI. Reprinted by permission of Shelley Posen.

Excerpt from "Terror in Black and White" by Robert L. Pitman and Susan J.
Chivers, reprinted from *Natural History,* December 1998/January 1999. Copyright
© 1998 by Natural History Magazine, Inc. Reprinted by permission of Natural
History Magazine.

Portions of this book first appeared in *Mother Jones* as "The Fate of the Ocean"
(March/April 2006), "The Thirteenth Tipping Point" (November/December
2006), and "Gone" (May/June 2007).

In memory of my father, Mark Stephens Whitty, who grew up sailing the wild Southern Ocean off Tasmania and who first took me offshore and taught me about wind, waves, and the power of coming about

Contents

Part One: Isla Rasa

1. The Very Air Miraculous   3
2. The River That Was Nowhere and Everywhere   14
3. Another Heaven   20
4. Hunger Island   26
5. The Ornament of the Body   31
6. One Hundred Days of Solitude   36
7. Whorls   43
8. The Unreefed World   51
9. The Epitome of Unrestrained Freedom   58
10. Mirage   64
11. Emotional Ecology   69
12. The Anti-Bodies of Quiet   73
13. Everything Is Already Brilliant   78

Part Two: The Underwater Rivers of the World

14. The Distant Geography of Water   87
15. The Ecumenical Sea   96

16. Deepwater Formation   103

17. The Tempest from the Eagle's Wings   107

18. One Meritorious Act   114

19. Jump Cut   121

20. Lament for the Thirty Million   131

21. All Time Is Now   138

22. Trophic Cascade   146

23. Bone Rafters   151

24. Soundsabers   159

25. Salting Down the Lean Missionary   167

26. The Existence of a World Previous to Ours   176

27. Reading God   184

28. Nemesis   195

29. The Inexplicable Waves   205

30. At the End of Hunger   209

Part Three: The Airborne Ocean

31. Serpent Cave   221

32. Black Mirror   227

Acknowledgments   237

Notes   239

PART ONE

~

ISLA RASA

O troupe of little vagrants of the world,
leave your footprints in my words.

—RABINDRANATH TAGORE
Stray Birds

1

The Very Air Miraculous

WORKING MY WAY back to the *casita*, the little stone hut we call home, I hop the boulders along the rocky shore-line, where relatively few of the 300,000 seabird inhabitants of Isla Rasa choose to nest. Even so, some Heermann's gulls—first-year breeders or latecomers or recluses—are nestled among these outermost rocks, napping on their eggs, confident of privacy, only to be rudely awakened by the approach of my dusty shoes. One after another, as if yanked on invisible strings, they burst into flight, webbed feet paddling, wings rowing backward, yowling *aow-aow-aow* in alarm and loosing globs of guano. I've been here for a month and still feel guilty about disturbing them.

It's late on an April day, and some strange trick of atmospherics is providing a hypnotic illusion. Perhaps it is "the very air miraculous" that John Steinbeck wrote of during his travels in these parts in 1940 with the marine biologist Ed Ricketts.[1] Here and now, sea and sky have merged into a pewter veil so thick the horizon line is erased. Yet the air is actually clear enough that flying birds cast black shadows on the icelike sea surface, invisible except for the shadows—a paradox of clarity and confusion created by a form of fog I've never seen before—not a whiteout in snow but a silverout in the subtropics, some 350 miles southeast of San Diego and 1,000 miles northwest of Mexico City.

I pause to study the effects of disorientation in a familiar landscape. At other times and places—in the weightlessness of underwater at night, for instance—I've lost enough sense of direction that vertigo upends my bearings. But here in the foggy heart of the Gulf of California, I'm anchored to Earth by gravity and affixed to the ocean by the sound of slap-happy waves. What's adrift here isn't the compass but time. As if a brontosaurus could unfurl its neck from beneath the waves or a pterodactyl flap ashore.

And then a prehistoric head does punch through the surface ten feet offshore, followed by the longitudinal ridges of a shell. It's a leatherback sea turtle, a creature straight out of prehistory, an inhabitant of the deep blue home for at least 110 million years, whose ancestors once shared the ocean with dinosaurs. From the length of the tail, this one appears to be a female. In the last pulse of light before darkness, she forms a perfect mirror-image twin with the surface: a two-headed turtle, jellyfish tentacles streaming from the corners of her mouths, like cellophane noodles in a silver broth.

The scene is transfixing. And not only because she's the biggest sea turtle I've ever seen, maybe six feet long and, I guess, a thousand pounds. Not only because leatherback turtles are rare or because it's the end of the day and living outdoors makes me perpetually hungry and right now even jellyfish soup sounds good. But because of the tableau of cause and effect rippling from it.

The turtle sculls at a leisurely rate, dipping her head, heaving the long tentacles into the air, swallowing, eyes closed to avoid the stings. As she chews the bell of the jellyfish underwater, the

curve of her leathery back—it's not really a shell at all—pitches
and yaws above the water line like a capsized dinghy. The sight is
intriguing enough that an elegant tern, another breeder on Isla
Rasa, detours from its flight path between feeding ground and
nesting colony to hover quizzically on butterfly wings. Other
terns notice this one's attentiveness, and a small flock coalesces
in the air, wings open, heads down, all eyes focused below, where
a school of damselfish bounces off the turtle's flippers, picking
her skin clean. A few Heermann's gulls congregate on the wet
rocks, slipping on algae, jostling for position, until one jumps
in and paddles counterclockwise around the turtle as she drifts
clockwise on the currents circling the island. The terns sideslip
through the air to keep up.

Just offshore, the entourage twirls into a floating raft of eared
grebes, thousands of small water birds paddling in close formation. They dive en masse when the turtle approaches, their
webbed feet stroking the surface before disappearing into the
ripples of their own making, then pop up in the same tight formation a hundred feet away. The flock will be here for only another day or two, en route to California's Salton Sea, three hundred miles to the north. It will fly there tonight or tomorrow
night, traveling nonstop, then make another nonstop flight to
California's Mono Lake, then split into smaller groups headed
for their birth lakes scattered through the Canadian Rockies and
northern Great Plains. Between here and there they'll cinch the
disparate bodies of water together as surely as threads in a necklace of blue beads.

A half-dozen Heermann's gulls accompany the sea turtle on
her perambulation of the current. They twirl on the paddles of
their feet and peck at the water, sampling strands of jellyfish goo.
Their species is adept at the thievery biologists call kleptoparasitism—stealing fish from other seabirds, particularly from the
gular (Latin *gula:* gullet) pouches of pelicans, though these gulls

aren't filching now because they don't eat jellyfish. Curiosity binds them to the turtle dining benevolently between their feet. Their persistent hopefulness binds them to the tableau.

From the deep waters beyond the grebes, a fin whale surfaces, sleek black back rolling forward a seemingly impossible length before the tiny dorsal fin scythes out of the water, travels the arc, then disappears below. The flukes stir a boiling cauldron of eddies without ever breaking the surface. A fishy mist of whale breath drifts my way.

Although darkness is falling, I follow the drift of the sea turtle on foot, retracing my steps along the shoreline to the easternmost valley of the island, where the sound of 30,000 breeding terns produces a collective voice as strident as a factory of metal parts gone haywire, broken steel and fan belts screaming. After a field season here, my hearing will never be the same. The terns creating this unlikely cacophony are pretty white birds bedecked with black crests and packed side by side within a bill's length of one another, in such tight formation that their nest scrapes assume the shape of perfect hexagons rimmed with guano and pebbles. This tessellated (Latin *tessellatus:* mosaic) pattern, like the cells of a honeycomb or the scutes (Latin *scutum:* shield) of a turtle's carapace, is one of nature's most efficient methods for packing space maximally.

The terns' work continues around the clock. As does the din. Even now in the dusk, thousands of terns are navigating through the congestion in the air and on the ground, hundreds of birds exiting the valley, hundreds more arriving, gliding in low, bills full of sardines, startling at the sight of the heaving turtle boat, their white wings pumping hard on the downstroke as they struggle to rise. Some careen close to my head, their distinctive *krrrrrk-krrrrrk* trills, the flutter from their wings, ghosting past my ears.

And then the leatherback drifts beyond where I can follow without trespassing on the tern colony I'm here to protect.

Anyway, it's nearly too dark now to see, so I turn toward home, climbing the familiar trail away from the tern valley, down into a gull valley, past the mountainous sand dunes formed of eons of powdered guano, then over another ridge and through another gull colony, before cresting the rocky ridge where the *casita* lies. Along the way I unnerve many gulls, who curse me in *ooh-ooh now-now* calls. As it does every spring evening, the sound of this island swells to fill the void of night.

In satellite images of Mexico's Gulf of California—that 700-mile-long finger of water splitting the peninsula of Baja California from the Mexican mainland—tiny Isla Rasa is invisible, lost to wave clutter and scale. The island is so minuscule that few maps bother to note it. Sea charts record an unnamed dot at 28°49′N, 112°59′W, in the part of the gulf known as the Midriff Islands region, a 60-mile-wide knuckle formed of numerous volcanic islands rearing from deep submarine canyons. Most of these islands are bigger, taller, more picturesque, more ecologically diverse, and far more enticing than Rasa to a human sailor plying these reaches. Isla Ángel de la Guarda, Rasa's largest neighbor, dominates its northern skyline with a reptilian spine of mountains 43 miles long and 4,300 feet high, painted in chameleon strata of red, ochre, and white that change hue with the season and the day.

In contrast, dun-colored Rasa barely rises above the waves, only 115 feet at its highest point and much of it lower than that. The word *rasa* means *flat* in Spanish—though the majority of English-language books and visitors mistakenly spell and pronounce the island's name "Raza," which means *breed* or *race,* an error that resonates faintly if you think of the island as the *race of birds.*

Isla Rasa presents such a humble profile that even from sea level many miss it. John Steinbeck, in the course of his travels with the biologist Ed Ricketts aboard the sardine boat the *Western Flyer,* never visited Isla Rasa or commented on it and most likely never noticed it as he sailed through the Midriff first to its west and then to its east on his voyage up and down the gulf.

Most travelers in these parts find little reason to stop here. Bereft of safe anchorage, fresh water, shade, or comestibles, the island lies forlornly empty nine months of the year, a tabula rasa but for a few raptors and corvids, some cryptic lizards, geckos, and invertebrates, plus countless tens of thousands of bird carcasses, mostly chicks, desiccating in sun and wind in a scene of postapocalyptic bleakness. Drought cruelly prunes Rasa's struggling garden, its dozen cardón cactuses, its crotches of prickly cholla, sour pitaya, and senita cactuses, along with saltbush and a few intertidal succulents. Most of the year the only animate presence is the wind, exhaling heat, coughing *chubascos* (thunderstorms), blustering low-pressure systems in from the north, and in all seasons spooling powdered guano and feathers into spindly dust devils.

Yet come February and March, vitality returns with a vengeance as the gulls and terns home in on their own unforgettable microcosm. The Heermann's gulls arrive from as far north as British Columbia, the elegant terns from as far south as Chile. This nondescript wafer of sun-blasted rock is the bull's-eye at the center of their lives, and within hours of their arrival they begin to fertilize the dead ground with their overwhelmingly dense, noisy, odoriferous, lusty, busy energies. Within days, Rasa houses an average of 2,700 seabirds per acre. When their eggs hatch, weeks later, the number doubles.

For the few frenetic months that they are here, Rasa's birds endow the island with a rare visibility, a transient topography forming and unforming in the sky as the air vibrates with hundreds of thousands of wings. From the sea, the island mushrooms into a jittery cloud visible for miles. In one season of four,

Isla Rasa is transformed into a curiosity that not even a hasty traveler through this land of oddities would wish to leave unexplored.

Some 95 percent of the world's population of Heermann's gulls breed on Rasa. The species is wedded to this island, partnered with it in a coevolutionary experiment between organic and inorganic nature. The birds' droppings prove a fertile boon to Rasa's thin soils. The birds' sooty gray plumage mimics Rasa's native lava rock, and their silvery white head and neck feathers match the guano-capped stone. The overall effect is of an animate, restless, flight-prone geology that blows into action and emits perpetual noise. The impression is further enlivened by the gulls' fire-engine-red bills tipped in black, as if a quarter million cheerful votives were perpetually lit against the island's glare.

Heermann's gulls are confident birds, striding upright and purposeful, hopping expertly between rocks with a flick of the wings, soaring buoyantly on long primary feathers. Yet compared to some of the giants of the family Laridae—the great black-backed gull of the eastern seaboard, for instance—Heermann's gulls are petite, with a small-boned, soft-voiced delicacy . . . though hundreds of thousands prove that even quietness multiplied can deafen.

Along with the gulls, as much as 97 percent of the world's population of elegant terns nest on Isla Rasa. Smaller and daintier than the gulls, they walk low to the ground, hunch-shouldered, strutting in fast-forward on seemingly motorized feet, orange bills outstretched as if following some urgent scent. Their plumage is strikingly white, highlighted with silvery wings. Their breeding plumage includes a rosy blush on the chest and a highly emotive black crest on the head that grows over the eyes, flipping and twitching with the expressiveness of an eyebrow. The ornithologist Arthur Cleveland Bent, in his *Life Histories of North*

American Birds, a twenty-one-volume set rolled out between 1919 and 1968, described elegant terns as "the most exquisite member of the charming group of sea swallows" and bemoaned the fact that their remote habitat prevented ornithologists from ever seeing one alive on the nest.[2]

A few thousand royal terns nest among the elegants on Isla Rasa. The two are difficult to distinguish except by careful observation. Given time, you might learn to see that the royals are slightly larger and more robust, their orange bills heavier and straighter, whereas the elegants' are decurved. The crests of the royals are also shorter and less flippity, and early in the breeding season they revert to the nonbreeding appearance of a black visor worn backward on a white head.

Both tern species arrive on Isla Rasa after the gulls have already established nesting territories on every usable inch of the island, usually early in March. The newcomers proceed to steal what they need, closing ranks under cover of darkness and dropping umbrellalike into a gull colony, flapping the spokes of their wings, jabbing the spokes of their bills. Come morning they've commandeered enough real estate to establish their own ternery, usually in one of the island's prime valleys, where the powdered guano lies as thick and soft as down pillows.

A fully packed gull colony forms an airy geometry, the birds spaced on two-foot centers. A tern colony nearly obscures the substrate, the birds packed into a seething, impenetrable sea of white polka-dotted with black, spiked with orange. In return for breeding inside what amounts to a giant gull colony, the terns are harassed every minute of every day by off-duty gulls or by subadult gulls patrolling the edges, fomenting squabbles, distracting the nesters, and stealing the terns' eggs. Nevertheless, the elegants invariably choose to nest beside larger gulls or terns, even when an opportunity to nest alone comes their way. On Isla Rasa the tradeoff boils down to this: better to live beside small predators who might eat a few of your eggs if their defenses help you survive the larger predators who might eat many more of your eggs.

Which is precisely the reason these small gulls and terns choose Isla Rasa over the neighboring high islands of the Midriff region. The flat topography allows for extremely dense nesting congregations, which provide the best defense for all three species from the island's larger predators: peregrine falcons, common ravens, and yellow-footed gulls. Flat Isla Rasa, a promising island to begin with, has been engineered over time more to the seabirds' liking—the rocky ridgelines growing softer and the valleys broader under a thickening blanket of guano. Thus the birds make the island, and the island makes the birds.

Each evening I return to the *casita* in near darkness, feeling my way by sound, weaving between the *aow-aow* calls of this gull and the *aow-aow* calls of that one. Eventually I catch sight of the weak light from our Coleman lantern spinning wildly through the air. Like so many of the things of civilization that we've brought with us, the lantern works poorly, clogged by the only fuel—aviation gas—we can get from our only contact here, a remote fishing village on the Baja Peninsula. Mónica is vigorously windmilling the lantern at the end of her straight arm, as if preparing to underarm a softball—her way of testing our working hypothesis that an intense kindling of oxygen will light the lamp, though mostly it just sputters gas, shadows, and frustration. Every night we have to restrain ourselves from pitching it into the sea.

Enriqueta is on the radio with Bahía de los Angeles, our lifeline for aviation fuel, tortillas, fresh produce, mail, and, most critically, water. It's a three-hour boat ride from the island to the village, and the trip aboard an open twenty-five-foot fishing skiff, or *panga*, can be made safely only in good weather. There's no Coast Guard here and no telephones, and consequently no fisherman lives long as a fool. Some years after this field season, a team of biologists from the University of California, Davis, will suffer five fatalities in this turbulent neighborhood where deep waters, high islands, and small boats intersect. We are persistently reminded of our mortality by three handmade crosses marking unremembered deaths in one of Isla Rasa's tiny valleys.

Before I set foot in the *casita* I hear alarm notes on the radio. The strange atmospherics are distorting the signal more than normal, stretching the voice from Bahía into a mournful oboe of concern. Have we seen a boat today? A small live-aboard cabin cruiser? A stranger appeared in the village this morning, a sun-burnt, dehydrated American who had trekked in over the rug-ged and waterless Sierra de San Borja after beaching his skiff far to the south. He had left his brother aboard their broken-down boat and gone for help in the tiny skiff, only to encounter cur-rents that redirected him wildly, as the currents in the Midriff tend to do, forcing him ashore and afoot in an unforgiving des-ert wilderness. Eventually he stumbled into Bahía de los Angeles and enlisted the help of a fisherman with a *panga* to return to his brother—only to find no trace of either the man or the boat.

Where was that? Enriqueta asks.

Anchored off the south coast of Isla Ángel de la Guarda.

From where we stand, the toothy monolith of this island is still visible in the dulling sunset to the northwest. We exchange glances. Everyone knows there's no safe anchorage there.

Mónica enters the radio room, the spitting lantern in hand. Its output is little more than one ghoulish candlepower, and its light warps the shadows cast on the stone walls by our hanging bird skins, blown eggshells, a half-dissected sea lion pup, the three of us. We stare, upset by this news of a distant stranger in distress.

Over the coming days the saga unfolds on the radio as other fishing *pangas* join the hunt and the few yachts in the Midriff are alerted. The brother charters a plane from the United States to fly a search mission. We have no boat, but we scan with binoculars many times from many viewpoints on the island. Yet as far as we know, nothing more is ever seen of the lost brother or the cruiser, and we are left restless and wondering. Only the birds and the fish, in their trembling clouds, know his fate.

2

The River That Was Nowhere and Everywhere

T HE TRUTH IS, the story I tell of a field season on Isla Rasa happened many years ago, in 1980. Enriqueta Velarde was a young woman closing in on her doctoral dissertation, on the breeding ecology of Heermann's gulls, for the Universidad Nacional Autónoma de México in Mexico City. Nowadays most of what we know of the science of this island and most of what I write here comes from Enriqueta's thirty years of continuous observations from one field season to the next on Isla Rasa.[1] Mónica was working on a study of elegant terns. I, an undergraduate wandering through academia after a teenage non sequitur in the world of professional theater, was working for Mónica.

Within nature's time frame and according to the slow clock of the deep blue home, my idyll on Rasa happened *less than yesterday*—though the elapsed time between *then* and *now* marks a cusp between eras in the Gulf of California, between a relatively pristine Edenic period (to paraphrase the Harvard biologist Edward O. Wilson)[2] and our present Underworld period, when the souls of so many species are crossing the waters to extinction.

Within these thirty years, science itself has changed. Many investigations have come to focus on the processes of extinction, some dissecting the five mass exterminations that have reshaped Earth dramatically in the past half-billion years, others uncovering the sixth great extinction currently underway. In these three seminal decades, our understanding has evolved to the point that we recognize how our excesses—habitat destruction, invasive species, pollution (including global-warming pollution), and human overpopulation—are converting this manmade extinction into a graver and more mortal threat than all others . . . one that seven of ten biologists believe poses a greater threat to human existence in the coming century than global warming, its contributor.[3]

We track the fate of life on Earth through the Red List, compiled by the International Union for the Conservation of Nature and Natural Resources (IUCN). Less than 3 percent, some 44,000, of Earth's known species have been scientifically assessed and categorized on the list's downward spiral from "Least Concern" to "Near Threatened" to "Vulnerable" to "Endangered" to "Critically Endangered" to "Extinct." Of these species, one in five mammals, one in eight birds, one in three amphibians, reptiles, corals, fish, conifers, and other cone-bearing trees are teetering toward extinction. Biodiversity, the living caul of a world perpetually rebirthing itself, is thinning dangerously.

In the *Mahabharata*, an epic Sanskrit poem of 100,000 couplets, there exists another Rasa, the River Rasā flowing through the Underworld, a Hindu River Styx bearing the souls of the newly dead from life. It was during my tenure on the desert island of Rasa that I first felt the griefs of the River Rasā wash my shores, as the vanished and nearly vanished travelers of the wild wafted by. The leatherback sea turtle (*Dermochelys coriacea*; Red List: Critically Endangered 2000)[4] visiting Rasa that spring evening in 1980 was one of 30,000 female leatherbacks nesting between Baja California and the Isthmus of Tehuantepec that year.

By 1996 less than 900 leatherbacks remained anywhere along the Pacific coast of Central and South America. Within that brief window of time, most of her long-lived race were killed in fishing gear or from ingesting garbage or pollution or the plundering of her species' eggs.

In the years following, I met other wayfarers clinging to the banks of life against the flow of Rasā: the world's smallest porpoise, the vaquita (*Phocoena sinus;* Red List: Critically Endangered 2008)—endemic to the Gulf of California and now reduced to 150 animals; the scalloped hammerhead shark (*Sphyrna lewini;* Red List: Near Threatened 2000)—once abundant here, now mostly memories. Each trip back to Isla Rasa and its oceanic neighborhood reinforced my impression of an inconsolable future . . . like the moment in the *Mahabharata* when the goddess Parvati tiptoes behind her husband Shiva, god of destruction, and playfully covers his eyes with her hands. Giggling, delighted at her game, she is instantaneously overwhelmed by Shiva's response—by the end of joy, by darkness shuttering the windows of the universe as the infinite blinks out and all hope fades.

Just like that.

I've often wondered: Did the leatherback I saw on Isla Rasa's shores that spring evening survive the next terrible decades? Since hatching, she would have meandered the 10,000-mile round trip many times between her feeding grounds off the western coast of South America near the Galápagos Islands to her likely birth and nesting grounds in the Midriff not far from Isla Rasa, fueling herself and her eggs primarily on jellyfish plus a smattering of sponges, squid, sea urchins, crustaceans, tunicates, fish, and floating seaweed.[5] Growing larger and less vulnerable with each passing year, she would have faced fewer dangers over the course of time until only the most oversized tiger sharks could harm her.

But then everything changed. In the late twentieth century, for the first time in the history of her species, a leatherback did not grow safer with age. Instead of invincibility, she faced mortal danger, as did all of her kind. In fact her risks increased so dramatically that by 2004 she gambled with a 50 percent chance of capture and drowning in fishing gear every single year. And that's assuming she survived all the other storms mounting in her path from the continued illegal capture of sea turtles for food, the growing risks of boat collisions and of the fatal ingestion of floating shopping bags, birthday balloons, rubber, tar, and oil. To make matters worse, the pressures facing her offspring multiplied too. At the onset of the twenty-first century, most of her eggs disappeared from egg poaching or from human construction and development on her nesting beaches. Many of those eggs that did survive to hatch suffered the lethal consequences of light pollution—the hatchlings seduced away from the moonlit sea toward streetlights and the headlights of cars.

In 1980, the point at which I begin to tell this story, the jelly-fish-eating leatherback sea turtle may have already encountered and successfully dodged many anthropogenic (Greek *anthropos:* man; Latin *genero:* begetting) troubles. During that field season on Isla Rasa the three of us were aware of the huge pressures mounting on her kind, and on all other living kinds. This knowledge was already changing our own small corner of inquiry, the science of ethology—animal behavior in the wild. Enriqueta and I talked often and wistfully of the classic field studies of Konrad Lorenz and Niko Tinbergen, who indulged in what seemed to us the delicious luxury of studying geese and gulls in the absence of humans. We understood, however, that our generation of scientists would be shaped by the inescapable responsibility of studying wild animals in wild settings forever diminished by our kind.

•   •   •

When I first visited the Gulf of California, the human population was so scarce and the wildlife so abundant and tame that many referred to this remarkable body of water and its archipelagoes as the *other* Galápagos. "We felt deeply the loneliness of this sea," wrote Steinbeck in 1940 of the waters not far from Isla Rasa. "No ships, no boats, no canoes, no little ranches on the shore nor villages. Now we would have welcomed a fishing Indian to come aboard and eat canned fruit salad, but this is a deserted sea."

Even in 1980, few Americans living beyond the American Southwest had ever heard of the Gulf of California. Most who had knew it as the Sea of Cortez, thanks to Steinbeck's *Log from the Sea of Cortez*—though many Mexicans dislike that name as much as they dislike the conquistador Hernán Cortés, for whom, it's said, only one statue stands in all of Mexico. For Mexicans, the gulf was their very own Wild West, a frontier for people who needed or wanted to flee the strictures of mainland life.

Yet it was not pristine when I arrived in 1980 and had not been since the closing of the first gates on the first dam on the Colorado River in 1907. For eons before that, the Colorado had scoured 1,500 miles of the drylands of the North American Southwest and transported countless cubic miles of pulverized rock downstream in seasonal blooms of silt. Millions of years of these flows sculpted the gulf's bathymetric map, building a sedimentary shelf one quarter the length of the Gulf of California 30,000 feet thick, reducing water depths to less than 600 feet in the northern half of the long finger of water. Most of the rock missing from the Grand Canyon came to rest under the waters of the Gulf of California.

The first dam was followed by a cascade of others, including the Glen Canyon, Hoover, Parker, Davis, Palo Verde Diversion, and Imperial dams. The All-American Canal, the largest of its kind on Earth, diverted so much Colorado River water to California's waterless Imperial Valley that it exceeded the flow rate of

New York's Hudson River, watering lush fields of fruits, vegetables, grains, and cotton—a crop requiring as much as four feet of rainfall a year. The reengineering of the Colorado inflicted a century of drought on the gulf, so upending its chemistry and hydrology that the nearly 2-million-acre Colorado River Delta, once a rival to the Nile and Indus river deltas as the richest desert estuaries on Earth, has become a withered ghost.

In 1922 Aldo Leopold personified the Colorado River Delta as "the river that was nowhere and everywhere, for he could not decide which of a hundred green lagoons offered the most pleasant and least speedy path to the gulf."[6] Today not a drop of Colorado River water reaches the Gulf of California, a development as monumental, in its way, as the tectonic opening of the gulf in the first place, some 4 million years ago. Consequently, this corner of the deep blue home, once a hotbed of biodiversity and prolific natural experimentation, grows simpler, quieter, and less alive.

3

Another Heaven

IN THE THIRTEENTH BOOK of the *Mahabharata*, after Parvati covers Shiva's eyes and all the cosmos goes dark, something incredible happens. A flame bursts from Shiva's forehead, and a third eye, a new sun, appears. It explodes upon the heavens, a jagged, pulsing, illuminating, bewildering lightning stroke. It immolates the mountains and forests where Parvati lives and fills the world with the scent of burning sandalwood. It stampedes herds of chital and sambar deer, flushes tigers, lions, leopards, civets, monkeys, otters, mongooses, wolves, jackals, sun bears, sloth bears, and elephants, all of whom flee to the protection of Shiva's regenerative shade. Parvati is so horrified at the destruction of her home that Shiva takes pity and breathes life back: the trees instantly flower, the animals trot home.

Afterward, when she asks why, he tells her that although her mountain was annihilated because of its proximity to his enormous power, his light revived the rest of a dying world. And so it is in Hindu belief that Shiva the Destroyer offers a counterweight to Brahma the Creator . . . in the same way that extinction on its routine and sedate course prunes evolution's weedy exuberance.

This afternoon, out of nowhere, as if the power has been cut, the cacophony of birds on Isla Rasa shuts off. Silence rushes in, followed by the sibilance of waves blooming heavily in our ears. The quiet is such a baffling guest that we lower our binoculars, stopwatches, field notebooks, and pencils, and we rise, swiveling on our heels, checking our backs. The factory of the island has closed shop. All the belts and screeching metal parts have seized up. To our amazement, the terns, packed in their blinding white mass, hunker low and wholly silent. The breeding Heermann's gulls (*Larus heermanni;* Red List: Near Threatened 2008) open their wings and hop a few inches into the air, transforming an entire valley of noisy, preening, courting, displaying, roosting, squabbling, panting birds into a single entity, a flying carpet of black and white and gray and red, flowing in Busby Berkeley choreography just above ground level to the water.

As the carpet settles onto the green linoleum of the ocean, fifty thousand wings snap closed, fanning a gust of air our way. The birds bob on the waves, silent, shoulders hunched, necks tucked. They're vigilant. They're frightened.

A dark force hurtles through the scene: a peregrine falcon (*Falco peregrinus;* Red List: Least Concern 2008), a steel-gray projectile as deadly as a boomerang, flipping from one tack to another on razor-sharp wings. Its course, at speeds in excess of 70 miles an hour, is so nimble and unpredictable, bouncing across three dimensions of air, that even from our viewpoint on solid ground the trampoline flight is dizzying. Tail flipping open, wings extending as if shot from an invisible cannon, it recedes to a distant dot overhead. Tail closed, wings cinched, it rounds an imperceptible apex and plummets, shoulders fanning out, conforming the body to the shape of a diamond. Sometimes it furls tight, hyperstreamlining at speeds between 100 and 200 miles an hour.

For whatever reason—inattention, sleep, foolish strategy, miscalculation—a few gulls on the rocky ridges did not abandon their nests with the others, though they do now, in panicked, flustered flight. The falcon follows, a missile mimicking their evasions with deadly ease. It shoots above them, slides under, rafts to the side. It skates past, brakes, doubles back, pulls up alongside. Yet it does not strike. To commit to violent bodily contact in high-speed flight demands decimal-perfect calculations. No fighter pilot has a tougher job. So the falcon, the cheetah of the air, endowed with explosive stamina, chases. Perhaps, in reality, it pursues for only a minute, though in the slow-motion world of an adrenaline surge—even for observers—it feels like ten. To the gulls it must seem an eternity, which it is, in its own way: the timeless cusp between this time and no time.

And then the geometry lines up. The falcon is chasing a gull, and before the human eye can register the change, it reverses course, climbs an airy ladder, and stoops upon a different gull. Striking feet first, talons splayed, the falcon knocks its prey into a somersault to the ground, out of sight behind a rock pile. A puff of powdered guano rises. The peregrine drops to its meal.

Later that afternoon we find the gull's wings, still attached by the wishbone to the pectoral girdle, plus the spine and skull. The legs and pelvis lie a few inches away. Nothing else remains but a pile of feathers, already disappearing into a feathered landscape.

After Shiva immolates Parvati's world, she asks him many questions, including why, since he possesses many luxurious abodes in heaven, he resides in a crematorium populated with vultures, strewn with funeral pyres and carrion, fat, blood, and entrails, echoing with the howls of jackals, surrounded by the hair and bones of the dead. *I always wander over the whole earth in search of a sacred spot,* says Shiva. *And I do not see any spot more sacred than the crematorium, shaded as it is by the branches of the banyan tree and adorned with torn garlands of flowers.*

So it is with the falcon aerie on the far side of the island, adorned with boulders and bones, heavily defended by sea cliffs and the terrifyingly aggressive displays of the peregrine falcon pair. It takes us weeks to find the nest site, even though Isla Rasa is minute and the cliffs are small.

The truth is, we inhabit only a tiny closet of this island house, plus a few hallways — the paths we've marked and cleared to minimize our impact on the birds. As the weeks and months pass, we see that those gulls unfortunate enough to be nesting beside our footpaths fare poorly: their eggs disappear, their chicks disappear, until most abandon their breeding efforts altogether. And so we walk literally on eggshells, lifting each leg, toes pointed, assessing each footfall in advance. The gulls' eggs and chicks are works of camouflage art produced with a splotchy paintbrush the colors of cement, dung, and slate, perfectly matching the environment. We depart from our marked paths rarely and only for good reason — in this case to determine whether the falcons, members of an endangered species, are residents of the island or visitors from across the sea.

I follow Enriqueta, her ever-present brimmed blue hat perched atop boy-short curls, as she advances into a monochrome landscape. This is her second field season here, and she moves with a precision and strength I'm still earning. Enriqueta is a paradox in timekeeping, at once nervous and trembly — her hands shake, her eyebrows quiver, she seems in a deerlike hurry to escape — yet also supremely relaxed, with the leisurely confidence of a tortoise browsing Isla Rasa, sampling this, nibbling that, metaphorically retracting her limbs when invisibility is required so that she somehow makes the birds forget she's here. Add to this that she's small, pretty, and very young-looking, with freckles scattered across her nose, and you might imagine that she is the junior member of the expedition.

On jaunts like this we stop often to scan with the near bionic extensions of our binoculars. Everything is informative. A distant lizard, a common raven (*Corvus corax;* Red List: Least Concern 2008), barrel-rolling through the air, a bouncing butterfly, the *casita* from this perspective, a daylight moon. Enriqueta shows me the waxy white flowers blossoming atop one of the island's giant cárdons (*Pachycereus pringlei;* Red List: Not Evaluated), a close relative of the saguaro cactus. The blossoms will last only twenty-four perfumed, musky hours.

When we discover the general location of the aerie, though not the aerie itself, we become the target of the falcons' terrifying dive-bombing runs—talons extended, wings furled, the machine guns of their *cak-cak-cak-cak-cak-cak* calls loosed inches from our heads. The male, smaller than his mate, has a higher voice. Together the pair produces the unnerving harmonic of a firing range.

We back away after only a few moments, time enough to understand something of the panic of birds who abandon their nest eggs, the promise of a distant future, in favor of an immediate one.

We search for the aerie mostly on Sundays, the day we award ourselves a respite from scientific duties, though science is everywhere here. We return again and again to the general area without finding the exact nest site.

One Sunday four men appear on the island unannounced, as all our visitors do. They're biologists from the United States Fish and Wildlife Service traveling the Midriff, surveying peregrine falcons. In 1980 the birds are in grave decline after decades of poisoning at the downstream end of the organochlorine pesticides like DDT, which have rendered their eggshells paper-thin. Clutches worldwide have been failing for years.

The traveling biologists are older than us, in good cheer at their paid adventure and full of questions about our island world. We're eager to share what we know. All of us are far enough from home that the wilderness has become our home, making us roommates of a kind. The men's surprise at finding us on this godforsaken outpost of dust and noise soon cedes to pleasure at the prospect of having an audience of (in all honesty) scantily clad young women with field notebooks. We in turn are impressed by their rock-climbing gear, live-aboard boat, and dinghy. We feel decidedly low-tech and studentlike, which we are, and hopeful that they might teach us to climb the walls of our island home.

At this moment, unbeknownst to any of us, peregrine falcons are rounding the nadir of their dive toward extinction. A few breeding pairs in North America, including, in all likelihood, this pair on Isla Rasa, have raised healthy chicks during these decades of failure. One of the earliest regions of peregrine falcon recovery in the world will be along the Baja California peninsula and on the islands of the Gulf of California, where the birds have been spared the worst—perhaps, ironically, because the Colorado River no longer delivers sediments, contaminated or otherwise, into their food web.

Two of the falcon biologists hike inland so as to approach the cliffs from behind. They wear climbing helmets to withstand the outrage of the parent falcons. Rappelling from the top, they peer into caves and crevices as the falcons scream in anger. They duck their heads, wave, smile, rappel further. They descend a line to the left. They descend a line to the right, describing in loud yells how tough a job falcon hunting is.

They work hard for more than an hour. Yet even with gear, persistence, and an audience, they can't find the aerie. So they drop to the ground, step out of their harnesses, coil their ropes, board their dinghy, and motor away. They have other falcons to visit, and we never get our climbing lesson.

4

Hunger Island

TWO YEARS LATER, Enriqueta makes a census of the entire island for falcon kills. Once a week for sixty-six days she walks every ridgeline and counts the carcasses. Her investigations reveal that during the 1982 breeding season the resident peregrines capture and consume forty Heermann's gulls, nine elegant terns (*Sterna elegans;* Red List: Near Threatened 2008), and one royal tern (*Sterna maxima;* Red List: Least Concern 2008). These kills amount to between 0.02 and 0.03 percent of the total nesting population of each species. In other words, just enough to support the needs of the falcons and their chicks without in any way threatening the future of the seabirds on the island and therefore the future of the falcons on the island.

The pressures of living with lethal roommates has forged a unique cultural response among Rasa's Heermann's gulls. In February, before egg laying begins, before they are tied to particular nest sites, the gulls perform mass flights—arriving on the island under cover of dusk, hundreds of thousands landing in a silent, eerie, locustlike cloud. Each morning before daybreak, they depart the same way, pouring offshore in a pyroclastic flow. Enriqueta suggests that the mass commuter flights to and from the island work to synchronize the gulls' nesting clocks and thus help safeguard them from the falcons.

In 1980 we're still kicking around a relatively new idea in science, proffered in the now classic 1971 article "Geometry for the Selfish Herd," wherein the evolutionary biologist W. D. Hamilton proposed that animals form flocks or herds not to ensure survival of the species but to facilitate survival of the individual—a revolutionary argument at the time.

The crowd offers salvation only to individuals within the crowd, said Hamilton, specifically those who can elbow their fellows out of the way to get to the heart of the crowd. He cited studies of black-headed gulls (*Larus ridibundus*; Red List: Least Concern 2008), showing that nesters on the margins of colonies failed to rear as many young, or even any young, since terrestrial predators picked off their chicks more often than those embedded in the crowd.

In 1980 the selfish herd is in evidence everywhere on Isla Rasa, though millennia of pressure have led to an understanding of sorts, the kind of harmony the ancient Greek philosopher Heraclitus described as opposing tensions, like that between the bow and the lyre. The peregrines tax but do not destroy the gulls, benefiting from the continuation of a gull-rich world. This realization is bolstered by another counterintuitive scientific revelation emerging during our tenure on Isla Rasa—that biodiversity thrives in the presence of both predators and competitors[1] and that the removal of either destabilizes the remaining species, potentially reducing biodiversity.

Thus Rasa's hunters cherry-pick the margins and tune and strengthen those gulls strong enough to hold the core, whose breeding successes fuel hungry falcons.

Every day the riddle of the aerie haunts us. The peregrine falcons fly over, terrorizing the island. The gulls pour offshore in silence.

The terns pancake flat on their nests in silence. Many days we find the characteristic cruciform skeletons of those who failed to pour or pancake fast enough.

Enriqueta and I return a few more times to look for the aerie. It's a pleasant quest that provides ample time to examine more lizards, scan the sky, study ravens, butterflies, the *casita*, the moon. We find other nests and other nesters, including a handful of yellow-footed gulls (*Larus livens;* Red List: Least Concern 2008) scattered erratically on lookout bluffs above the sea. We find pairs of American oystercatchers (*Haematopus palliatus;* Red List: Least Concern 2008), though never their nest scrapes, hidden expertly between shoreline rocks.

Our circumnavigation takes us past the nests of the island's ospreys (*Pandion haliaetus;* Red List: Least Concern 2008), perched on prominent bluffs overlooking the sea. One nest is inactive. The other, a monstrously shaggy six-foot-tall affair, crowns the northwesternmost point of the island and punctuates our view from the *casita*. It is composed of an overflow of skeletons, cactuses, driftwood, mummified seabirds, leatherized kelp, old sneakers, flip-flops, baseball caps, hemp fishing nets, and other flotsam and jetsam. Deconstructed and tallied, this nest would doubtless prove an interesting inventory of the flow of materials through the Midriff over a span of decades, maybe longer. The osprey pair tending this nest are raising two chicks on a diet of coronet fish, judging from the pile of long-jawed skulls scattered on the ground.

Not far from this nest are other tall structures of similar size—Rasa's mysterious stone cairns, mostly conical sculptures roughly five feet high and seven feet in diameter at the base. The island hosts more than 11,000 cairns, by one count, along with extensive rock walls that punctuate the hills and ridgelines like frozen armies of stone. These perplexing troops dominate the landscape in such multitudes that they form Rasa's most inescapable feature aside from its birds, which so overwhelm the overwhelming cairns that many visitors fail to even mention the cairns' existence.

No one knows the genesis of these rock structures, which pave the island like the broken-down architecture of a vanished society, although they are probably the remnants of the extensive guano-mining operations that once visited or plagued Isla Rasa, depending on your point of view. Rasa's seabird guano, rich in ammonia and nitrates, was the prized fertilizer of a nineteenth-century green revolution. Beginning in 1868, 10,000 tons of guano a year were mined from the island and delivered to Germany. Later entrepreneurs extracted diminishing returns until the 1910s. One or all of these operations likely constructed the cairns.

The survey ship USS *Narragansett*, visiting Rasa in the 1870s, reported on the mining operations then underway: "The surface guano is collected in the form of dust and shipped in bags. The layer succeeding it is composed of 'clinkers,' which require crushing before using. The 'clinkers' are richer in phosphates than the pulverized guano and are more easily gathered and shipped."[2]

We who also labor on this island wonder about the effects of this mining on the birds. The ornithologist Griffing Bancroft, who visited Isla Rasa in the 1920s, claimed to have found evidence of old abandoned burrows of Manx shearwaters (*Puffinus puffinus;* Red List: Least Concern 2008).[3] In 1980 the three of us find the dead body of what we believe to be a Manx shearwater, trapped in a thicket of cholla cactuses. (Today these birds would likely be considered a different species within the Manx shearwater complex, perhaps the black-vented shearwater [*Puffinus opisthomelas;* Red List: Near Threatened 2008].) Could it be that some member of the superspecies *Puffinus puffinus* nested on Isla Rasa prior to the cairn-building era and that its population was destroyed by the mining process?

For all the stories we come to know, however obliquely, thousands more are lost forever.

One day we find the peregrine falcon aerie not far from where the Fish and Wildlife Service men were searching. It lies on a ledge below an overhanging ledge, a site nearly impossible to discern from above or below and inaccessible from any direction without wings. It's a cryptic masterpiece. Chances are it's been in exactly this location for generations, since we know that some aeries on European islands have been continuously occupied for 350 years, and probably longer. Falcons engineer the simplest of nest scrapes, saucer-shaped depressions in debris on rocky ledges. Yet their residency produces a buildup of droppings on and below the ledges that can reach epic proportions. An analysis of one pile in Australia revealed a staggering 16,000 years of habitation.[4]

Half scrambling, half climbing, we gain vantage into the secret enclave. Paradoxically, we can see inside only from far away and only when a cloud covers the sun, erasing the contrast between light and shadow. A downy chick stares back with black, unblinking eyes, already intimidating. A month from now, enrolled in flight school, it will terrify the birds of Rasa on an hourly basis, flushing gulls first from one valley and then another, the mass flights flowing in silence to the waves to bob and wait, fly back, be flushed again. It's an exhausting, uncompromising cycle, during which the fledgling falcon learns something of its power—physical and psychological—over its world.

Meanwhile its parents dive-bomb us. We cling to the cliff on taut toes and fingers, suspended by awe.

5

The Ornament of the Body

IN SANSKRIT, the classical Latin of India, the word *rasa* refers to a cosmic dance performed by the dairymaids called *gopis*. In the *rasa* dance, the *gopis* revolve around the still center of Krishna, the divine incarnation of all their desires, master of the sixty-four arts of love, and renowned paramour of many *gopis*.

You might expect such a dance to be a frenzy. Yet it is so contained, stately, silent, and slow that in the course of its long unfolding it elevates time from its rutted track to an infinite fetch. At its core, the *gopis' rasa* is said to emulate the continuous, unbroken orbits of the planets around the sun—the original incarnation of gravitational desire. Krishna's favorite *gopi*, Radha, sings to him: *As wing to bird, water to fish, life to the living—so you to me.*

Apropos of its name, Isla Rasa resonates with its own *rasas* each spring. Amid the complex social dynamics of friends, foes, neutrals, and allies, amid the time-consuming duties of feeding, preening, sleeping, and bathing, the gulls and terns arrive with one sole purpose: to breed. To succeed, each bird needs to find a mate, either a new one or its life mate from seasons past—the one who will, in Radha's words, *put the ornament of your body upon me.*

But bonding requires collateral. Krishna played a bamboo flute to enchant Radha and win a first kiss. The terns offer a sardine and seal the bond with a dance.

I have a regular seat on the island, a flattish rock at the end of the small trail that runs laterally along the hillside on the western edge of the tern colony. Small clusters of cholla grow nearby, some of which "jump" to my backside on absent-minded or exciting occasions. I spend many an hour on this flattish rock staring down at the tern colony, surrounded by gulls, shifting from one buttock to the other, juggling binoculars, notebook, clicker. Along with my regular scientific chores, which amount to collecting predetermined data sets designed to address all kinds of questions, from aggression to colony growth to kleptoparasitism, I simply observe.

For a long time I don't see much. The tern colony is overwhelmingly chaotic. Imagine John F. Kennedy International Airport at Christmastime. Jets are coming and going. People are coming and going in crowds so dense it's hard to differentiate individuals. All are carrying luggage, trying to stay together in family clusters. There's a lot of jostling and shoving, testy exchanges, rudeness, occasional kindness, flirting. Everyone's shouting to be heard. Thieves are patrolling the edges, trying to steal luggage and children. Imagine JFK Airport for a species other than your own, whose customs you're not familiar with, speaking a language you don't understand, inhabiting a challenging landscape full of plants as sharp as swords and clouded with gnats that lap at the tears in your eyes.

Because the tern colony is so crowded, much courtship and breeding take place on its edges among clusters of birds, known as clubs. The endearing qualities of elegant terns are most evident in their clubs: the little birds strutting on motorized feet, pairs circling each other, black crests flipping up and down, shoulders extended, wings dragging slightly on the ground. Other groups, known as forums, arise when as many as eight birds face each other in a tight-knit circle, flicking their heads and crests in a staccato rhythm.

In the clubs and among the forums the birds attempt to co-ordinate their activities, attuning themselves to one another psychologically and physiologically. At Club Sardine, for instance, a club or parts of a club might decide to fly off together in the safety of a flock to forage offshore. At Club Love, terns might look for old mates (elegant terns are probably monogamous) or find new ones. At Club Newlywed, pairs synchronize their breeding cycles, choose nest sites, perform courtship formalities. In most of the clubs, some or all of these activities are going on simultaneously.

Seated on my flattish rock, I'm a voyeur with a camera, trying to document each stage of elegant tern courtship. The choreography is as formal and ritualized as a medieval court dance, designed to synchronize not only the intentions of a single breeding pair but the intentions of all the breeding pairs on the island, since the colony's joint defenses are most effective when all birds are busy doing the same thing at the same time. This coordinated schedule makes it easier for me to learn to see what's happening, to focus on one activity at a time.

In the early phase of tern courtship I chase (with my lens) many a hopeful male suitor. They are easy to pick out of the crowd as they parade through the club, walking tall, neck stretched, crest raised, wings extended and drooping. Each carries a small silver offering, a sardine, draped delicately in the tweezers of his bill, held as close to the tip as possible, as if to say: *It's not really mine, it's for you.* He gives repeated *cher-wit, cher-wit* calls, flicking his head about once a second to make sure that his edible bauble dances enticingly in the sun.

A hopeful male performs his fish-offering display as he pursues a female performing the begging display, a demonstration common among female gulls and terns. She hunkers low, legs bent, belly almost on the ground, neck outstretched parallel to the ground, ruffling her neck, shoulder, back, and wing feathers. This quickly leads to a feeding chase, the male with sardine running after her, the hunched female running away—though not too far and not for too long. If the moment is not right or the partner is not right, she'll abandon the footrace and launch into flight. End of date. If she likes him, she'll stop and accept his sardine. The engagement ring.

Before they can wed, however, they need to assess each other in their preferred realm, the air, where they cement their bond in a series of paired courtship dances—displays so ethereally beautiful they never fail to lure me from the ground, if only with my eyes. As soon as the female accepts a fish, the male launches into the air. She follows on quick wings, flapping hard to position herself below him, and the two soar, one above the other, wings forming perfect white V's against the blue sky. The male calls *kit-a-leek* as they glide. After a few of these glides, the pair ascends side by side, white arrows climbing, gaining so much altitude I have to squint to see them. Some pairs disappear beyond my sight, then fall together like Siamese shooting stars, perhaps nearly touching, flaring at the bottom into another series of glides.

Anchored to my flattish rock, I watch.

If two beings can pull off this level of harmonization, they can afford to commit to the enormous full-time investment of time and energy needed to defend a nest, produce an egg, incubate an egg, brood a chick, feed a chick, and migrate with a dependent chick for six or more months. Only a few months after that, they'll start again, usually with the same mate, after the same soaring courtship flights—perhaps the only moments of lightness in their hard-working lives.

Despite all the obstacles to love and success, most birds succeed. Day and night for weeks the island is a single-minded mass of copulation. Males hop on females and balance precariously between their wings as the two rub their backsides together, tails cranked to the side. The operation looks and is precarious, since there's no anchor, no penis—an organ missing from the anatomy of virtually all species of birds (the exceptions are the large flightless birds like ostriches, some ducks, gallinaceous birds like turkeys and grouse, and tinamous).

Instead, the external genitalia of most birds is a unisex cloaca (Latin: canal), a common passageway for bodily wastes and gametes (eggs and sperm). Rubbing two cloacae together while two birds balance on one bird's two feet is challenging, often comical. Nevertheless, the gulls and terns accomplish it, again and again, around the clock, for the duration of the breeding season. Biologists call it the cloacal kiss.

6

One Hundred Days of Solitude

SURROUNDED BY BREEDERS, we three women are celibate, having left our mates behind in worlds so different that there will be no way back for any of us at the end of this field season. We sometimes feel our partners' absences keenly but mostly in a theatrical way that heightens the adventure we've embarked upon. Without doubt, it's better here without them. This momentary hiatus in our own otherwise inescapable programming is our first opportunity since puberty to soar our own way.

Since there's no one to flirt with, we abandon grooming routines, forgoing the tasks of shaving legs and underarms, plucking eyebrows. Alone, we begin to reinvent ourselves, dabbling with the laboratory-strength hydrogen peroxide we use to prepare bird skins. Suddenly we're an island of blonds with hairy (blond) legs. Since we wash only in salt water, our hair gets bushier. We morph from pale to supertanned. In no time at all we've gone completely feral. Oddly, in the absence of native English speakers, I begin to speak a distinctly Mexican-accented English, while Enriqueta and Mónica develop a more American-accented English. The few visitors who make it to the island have no idea what to make of us.

Meanwhile, we experiment with the really interesting enterprise of learning to see in another language. That is, to learn to see in Gull or Tern or Falcon. They're different species, of course, as well as different intelligences, being altogether smarter than us in matters of flight, fishing, hunting, migrating, orienteering, outdoor survival, and marriage.

In the Rig-Veda, a collection of Vedic hymns from ancient India, *rasa* also refers to living liquid, like the juice of fruit or the sap of trees—an essential fluid, sometimes called the *semen feminile*. It suggests the liquid core—the broth—of anything: the milk of the breast, the marrow of bones, the juice of sugar cane, *ghee* (clarified butter), the vital juices of the body.

Like all liquids, this *rasa* flows. In its metaphysical form, it pours across boundaries we normally consider inviolate: bodies, species, ecosystems. Isla Rasa affords a portal into the mysteries of this current. Water flows onshore in the form of fish carried to birds' nests. Land flows offshore in the form of falling guano.

Enriqueta, Mónica, and I find our place in the flow, jettisoning scholarly physiques and congested brains. The island sucks the other world out of us. My journal records our transformation to brown outdoorswomen with burgeoning amnesia and muscular questions ranging well beyond science. We shed our prior lives swiftly but for a strange smattering of technology: binoculars, dissecting scopes, mist nets, a cassette deck with tapes of Mercedes Sosa, Patty Smith, Talking Heads, Elvis Costello. Oddly for a reader, I've brought only one book, Gabriel García Márquez's *One Hundred Years of Solitude*. I didn't know it when I added it to my gear pile, but it's the only book I'll need here, one I'll savor page by page by sputtering lantern light in the course of a hundred nights.

We're aware of another famous Spanish novel, as beloved in the sixteenth century as García Márquez is today—Garcí Rodríguez de Montalvo's *The Exploits of Esplandián,* home of another fabled kingdom, one ruled by Queen Califia, for whom Alta and Baja California were named:

> To the right of the Indies, there was an island called California, the nearest thing to an earthly paradise, populated by black women, with not a man among them, who lived almost like Amazons. These women were endowed with strong, brave bodies, passionate hearts, and great strength. The island itself had steeper rocks and more rugged cliffs than any other in the world. Their weapons were made entirely of gold, and so were the harnesses of the wild beasts on which they rode after having broken them in.[1]

Of course, we don't ride wild beasts and there is no gold. Still, we jokingly refer to Rasa as Isla de las Amazonas as we become adept at uncivilization, learning to make do without many things that would be useful in our deserted sea but that we forgot to bring, including any kind of cooking pan (we prepare everything in a coffee percolator), chairs (we sit on rocks), warm clothing (we toughen), a boat. For a time we are disheartened and possibly poisoned by our fifty-gallon plastic water barrel, which previously stored aviation gas and is still infused with that taste and scent. The truth is, we're as good as marooned on a desert island without a palm tree, and little by little we unclench all our roots.

Most days we hear no sounds of other human beings. The island is deeply buffered from the faraway world, lying miles from any flight and most boat paths. We quickly become attuned by the finest of the tiny stereocilia (Greek *stereos:* hard; Latin *cilium:*

eyelash) in our inner ears to any noise from the outside, and long before we are able to see a boat on the horizon, we receive the rumble of its engine in our bones, the rising and falling whine of an outboard, the throaty gurgle of an inboard.

For as long as we are aware of a boat, we're held in its thrall, unable to process any other task or thought. It's a bewitchment that provides a visceral link to the lives of our forebears, cocooned in their intimate clans, wary of unknowns. We act on their ancient programming when we target approaching boats with binoculars and wait, suspended between curiosity and suspicion. If a vessel appears to be headed our way and likely to make landfall, we converge on the landing point, a guardian threesome, geared for peace, able for battle. The whole process, from our first awareness to the strangers' arrival, often absorbs an hour of our time.

Invariably, first-time visitors are surprised to find us here. Three young women, scantily clad in the shortest of shorts, skimpy tank tops, braless. We are the younger benefactors of the feminist wave of the 1960s and defiant of our right to immodesty. Our female visitors, mostly older American and European tourists and yachtswomen, display a maternal blend of pride and envy shadowed by doubt. Our older male American visitors often take us aside, literally by the arm, to ask, sotto voce, whether we have a firearm with us. We laugh. We're young, invincible. We can't conceive of tragedy.

This morning a *panga* is circumnavigating the island faster than we can run to keep up with it. It's here, it's there. We suspect the mission: fishermen, *pescadores,* in pursuit of a boatload of eggs. They've traveled to this remote island not because there's a shortage of laying hens in their village but because the seabird eggs of Isla Rasa are reputed to be aphrodisiacs. I mutter as I chase. Mexico in 1980, with one of the highest birthrates in the world, doesn't need aphrodisiacs.

The *pescadores* know we're after them. They feint this way and that, counting on their ability to get in and out of a colony faster than we can. We sweat and stumble and bleed from the punctures of cholla. The gulls along our paths are incensed. We're incensed.

At one point, from the far side of the island, I see a white mushroom cloud blossom above the tern colony. The men have landed and flushed the entire colony off their nests and into the air. They'll smash the eggs and return a few days hence for the newly laid ones. Or they'll abscond with what they can today and hope that at least some of their "catch" is freshly laid.

In the late nineteenth century and for most of the twentieth, egg collectors were free to move in, shatter a whole valley of eggs, set up camp, and wait for fresh ones to be laid. As early as 1922, visiting scientists realized the threat to the future of the island's birds. Griffing Bancroft found, anchored off Rasa, "little flotillas of three or four tiny sailboats filled with eggs to be sold as food in Santa Rosalía," and in 1928 he discovered that the entire breeding season for both gulls and terns had been effectively wiped out.

Commercial egging continued until 1964, when Mexico designated Rasa as the nation's first bird reserve. A year later the biologist R. J. A. Barreto estimated that only 25,000 birds, including chicks, were present on all of Isla Rasa—fewer birds than inhabit the small valley behind our *casita* in 1980. Protection came not a moment too soon.

So we chase the fishermen, and when we stumble over the last ridge to the tern colony we come face to face with an apocalypse. The terns are off their nests, tornadoes of white swirling through the air with a sound like fighter jets launching from a carrier. The *pescadores,* three of them, are buried in the frenzy, invisible except for the angriest clot of birds converging around them. Now and again a window opens in the feathers and we glimpse a man's leg, a hand.

The days of commercial egging are over, yet there still is, and may always be, the occasional local fishermen in pursuit of an easy peso or the difficult heart of a winsome *señorita*. We wish to educate and dissuade our insolvent or lovelorn poachers—though we debate exactly how to do that. Should we charge into the colony and drag them out? Thankfully, we don't need to. They see us and exit of their own accord, hounded by terns and gulls and dripping with guano. We meet them on the seaward edge of the colony, beside their *panga*, three men facing off against three women, two sides with different agendas. Around us tens of thousands of terns scream their way back to their nests, and thousands upon thousands of gulls capitalize on the pandemonium to consume the undefended terns' eggs. We stare at each other, searching for a bridge.

Despite the fears of our American visitors, these fishermen, like most of the Mexicans we meet on this island, are so politely formal as to meet some long-forgotten definition of chivalry. Enriqueta is calm, though her eyebrows quiver and her hands shake. Mónica is stern, a disapproving *madre*. I'm jumpy and mad and ready to fight. Enriqueta and Mónica explain the realities of the island, the law of the land. They trade back and forth, good cop, bad cop. The fishermen stare at the ground, shuffle their feet. In the end they decide that the best way to build a bridge is by bestowing upon us the gift of three fresh lobsters from the bottom of their *panga*. We are agreeable. Hands are shaken all around. The men wave a friendly goodbye. We are *Amazonas* after all.

Without doubt, these are the best lobsters we've ever eaten, and for me the last. A year from now, scuba diving with a film crew in the gulf not far from Isla Rasa, I encounter the unforgettable, albeit fleeting, drama of a spiny lobster (*Panulirus interruptus;* Red List: Not Evaluated) defending its cave home from a much larger, lobster-eating horn shark (*Heterodontus francisci;* Red List: Data Deficient 2006). The shark, swimming away from the human divers, beelines toward the cave and tries to enter, but the lobster stands firm in the mouth of its cave, antennae twirling. The shark is bigger, tougher, and determined to nose in, but the lobster cracks the whip of its antennae onto the shark's snout and taps—*tap, tap, tap, tap, tap*—a Morse code of crustacean resolve. The standoff lasts a tense moment, the energy palpable as the two lean into each other. Then the shark backs out and swims away.

My last lobster meal is accompanied by a cucumber salad, fresh off a resupply boat from Bahía de los Angeles. We toast our successful job as wardens with cups of wine poured from a cheap boxed rosé endowed by visiting Americans who took pity on our alcohol-less state. Meanwhile, three valleys away, the tern colony is recoiling from the intrusions of the fishermen as the gulls crash the gates. Within hours, the single protoplasm of white tern feathers and flaming orange bills splits, amoebalike, into a pair of daughter tern colonies separated by a moving stream of Heermann's gulls patrolling a new path of their own making, a new inroad against the selfish herd.

A couple of weeks later a different threesome of lobster fishermen repeats the encroachment into the ternery. We confront them too, but because they prove altogether less friendly, we feel altogether less confident and therefore act less authoritatively. They depart, unsmiling, after a tense standoff, without offering any lobsters, which we might not have accepted anyway.

Afterward, the twin tern colonies maintain their borders but are noticeably thinned of nests. For many terns, the breeding season has been derailed. The blinding white of the birds is replaced by the blinding light reflected off countless broken eggshells, and the whole island grows quieter.

7

Whorls

MANY NIGHTS, suspended in my hammock, I dream that the laughing *ah-ah-ah-ah* calls of the gulls nesting under the walls of the *casita* are the voices of children speaking a language I can't understand.

This night we're awake and sleepless from cold. A low-pressure system blowing down from the north barrels through the glassless windows and doorless doors of our hut, a jet stream on its way south. We huddle deep in sleeping bags, fully clothed, shivering. We skipped dinner because we couldn't keep the Coleman stove lit in the ripping air and couldn't stay warm in the kitchen. Now, tired and hungry, we listen as the mortar holding the stones of the *casita* together pops loose under the air gun of the wind and ricochets against the walls.

For the first time since our arrival we can't hear the birds. Their voices, normally cacophonous, are lost behind the shriek of the gale and the clanging, thudding, clunking, and hammering of our gear hanging against the walls. None of us has ever been in the presence of such a wind. It's too loud to speak, so we lie awake in the company of our observations.

My hammock rocks like a dismasted sailboat, bumping ashore on the shoals of the rock walls. Enriqueta's cot, wedged into a corner of the room, collects drifts of powdered guano. At three o'clock in the dead of night with the wind still building, we suddenly wonder about the fate of our twenty-foot-tall radio antenna, rigged on two-by-fours and precariously tethered to the hut with guy wires and rusty nails. Peering into the moonlight, we see the antenna leaning hard, its rock anchors within inches of sliding off the roof.

We're concerned about our communications but more worried about the fate of the gulls sitting on eggs below. Their behavioral repertoire does not include abandoning their nests in the dark of night because a large antenna is in danger of falling, so Enriqueta and I lurch into the wind, heads down, eyes squeezed against the airborne grit. We lean against the antenna with all our strength, but try as we might we can't straighten it. We settle for shouting at Mónica inside to use strips of mosquito netting to tie the antenna against the none-too-secure warped wooden window frame of our bedroom. Since we can't hear each other above the keening in the guy wires, we revert to pantomime.

We fix the antenna as best we can and return to a wind-rattled insomnolence in our camp beds. By dawn the wind calms to a Force 6 on the Beaufort scale, euphemistically referred to as a "strong breeze." It's blowing hard, yet not so hard as to mask the welcome return of bird dissonance. Offshore, the water heaves with swells and somersaulting whitecaps. Amazingly, the antenna still leans.

Our supply-boat driver from Bahía de los Angeles, a tiny, wizened, reclusive man nicknamed, for reasons unknown to us, Pinguino (the Penguin), has been stuck on the island for two nights because of the wind, sleeping beneath the thwarts of his *panga*. He stands beside the antenna now, calling softly, his back turned away from the embarrassing proximity of sleeping *señoritas,* his hands full of dead birds.

Over the coming days we find mangled dead seabirds all over the island, necks broken, wings broken, legs broken. Because we're preparing bird skins, we bring these victims back to the *casita.* Beneath their soft feathers and skin, the glistening muscle fibers that powered their flight are blackened with bruises.

In the world on the other side of the tropic of civilization, people assume that for birds the wind is a benevolent ally. But those of us at work in the deserted sea understand that birds are sailors too, their boats painfully vulnerable to wind swells, wind troughs, rogue air waves, rip air tides, and atmospheric tsunamis. When things go wrong, these fliers also wreck on distant or near shores or drift powerlessly on the invisible currents of the River Rasā.

We dismantle their limp carcasses with scalpels and refill their empty skins with gauze, heaping the innards in pyramids of shiny purple and red organs as pretty as multicolored beans. The intimacy of the task, requiring many hours of work and repeated many times over many days, brings us closer to our fellows on this island. We see the causes of their deaths (the broken bones), the causes of what would have been their deaths (the tumors), the symptoms of their dying (the empty stomachs), the mysteries of their seemingly excellent health. Like pathologists, we live with their inner stories day after day, absorbing their lives and deaths through the whorls of our fingerprints.

The indigenous Seri people of the eastern shores of the Gulf of California knew these fierce northwesterly blows that bedevil birds and sailors and were aware of them a day or two before the winds arrived in the Midriff. Gales to the northwest raised dust storms on the distant horizon, foretelling bad weather. The dust spiraled upward from the vast Laguna Salada, the salt pan the Seri called Caailipoláacoj, the Great Black Dry Lake.[1] Historically this was less a lake than an occasional sea, since it lay at the uppermost fetch of the Colorado River Delta and was flooded on extreme tides, which deposited marine fish and squid in an ephemerally transformed desert.

The same tempest that blows us so hard and carries some of Rasa's living birds to their death also delivers wind-bedraggled newcomers: a Le Conte's thrasher, a Costa's hummingbird, a bloom of butterflies, Mexican fritillaries. Blown from the north, they've made landfall as best they could. The neighboring islands are doubtless decorated with hijacked spores too, the drifters who may or may not get home again. Some will find new residences in territories already populated by their own kind. A rare few will populate a novel world. Others will die over the water or fall prey to unfamiliar predators. I note their arrivals in my journal and hope their subsequent disappearances are signs of success.

The water driven by the winds also carries spores, though more slowly, since a surface current moves at only about 3 percent of the velocity of the wind. Nevertheless, for days afterward we stumble on the oceanic aftereffects of the big blow—the limp, crinkly kelps and algae that have come to rest on the windward shore, their holdfasts gripping rocks ripped from the sea floor. They wash up in piles like the discarded clothes of a party of skinny-dippers, to be visited by Sally Lightfoot crabs—the island's meticulous housekeepers, who fold and unfold their rags with red enamelware legs, rifling their pockets for comestibles.

The Midriff, with its bare-knuckle islands and deep underwater canyons, produces a surfeit of wind-driven currents. From the shore of Isla Rasa, these embedded rivers in the sea are visible most days. Upwellings appear as boiling whirlpools stamped with huge, flat circular footprints. Downwellings show as oily calms in the midst of riffles of wind. Langmuir circulations form during days of light, steady winds. Their churning rearranges the surface into parallel sets of windrows, those curious debris lines of foam, seaweed, and flotsam that span miles of the surface, each windrow marking the center of a pair of Langmuir circulations, whose slow-turning, counter-rotating vortices roll continuously through the top sixty feet of ocean like enormous invisible waterwheels.

Upwellings are the most biologically productive of all currents: vertical conveyer belts rising from the abyss to the surface, bearing the sunken components of dead plants and animals in the form of dissolved organic matter. This rich broth is destined to fertilize the phytoplankton in the sunlit zone, whence much of the dissolved organic matter originally came. Upwellings can occur anywhere, including midocean, though the superproductive ones develop along coastlines, where prevailing winds blow parallel to the shore, pushing the surface waters ahead of them.

But because the flow of wind-driven water is also influenced by the Coriolis effect (a spinoff of Earth's rotation), all currents are deflected to the right of the wind in the Northern Hemisphere and to the left in the South. And because the ocean is stratified into density layers, the Coriolis effect redirects these tiers too. The surface-driven current, spun by the Coriolis effect, may tug on the layer below it, which tugs on the layer below it, in successively lesser degrees of strength. The end result is a downward whorl known as an Ekman spiral, which may corkscrew through the water miles below its originator, the wind. The net work of all the layers is known as the Ekman transport, which has the theoretical power to deflect water 90 degrees off the wind.

Consequently, breezes blowing parallel to a coastline actually result in water flowing offshore at a right angle. The offshore flow is then replaced by water rising from the deep, bearing its Miracle-Gro of nutrients destined to feed blooms of phytoplankton, which feed the zooplankton, who feed the sardines, who feed the tuna, dolphins, whales, and seabirds. The plankters that escape being eaten sink to the sea floor upon death, their ghosts eventually resurrected at the surface to fertilize their own kind, maybe their own kin, some generations hence. Thus the deep blue home recycles matter and energy with impeccable efficiency.

The physics of the Ekman spiral were deconstructed in the early twentieth century by the Swedish mathematician Vagn Walfrid Ekman, who was investigating a problem posed by none other than the great Arctic scientist and explorer Fridtjof Nansen. In the course of three years adrift and afoot on the Arctic ice, Nansen wondered about the curious 20- to 40-degree deflection of icebergs in relation to the wind. Upon his return to civilization, he suggested that young Ekman explore the mystery. Thus the spore of a question born during Nansen's epic Arctic circuit made landfall in Ekman's mind, eventually to blossom into a new understanding of fluid dynamics that today helps us predict the behavior not just of oceans but of weather and of nebulae in interstellar space.

The Gulf of California, with its many coastlines along numerous islands and two parallel shores, is turbulent with upwellings, downwellings, Ekman spirals, and Langmuir circulations—powerfully busy waters rushing in opposing directions on a three-dimensional scale. The effects are intensified by the gulf's thirty-foot tides, among the most extreme on Earth and the offspring of the region's unique topography and bathymetry.

Most large oceanic bodies circulate tidal energy around a no-tide center known to oceanographers as an amphidromic point. But the gulf is too narrow for an amphidromic point to form. Instead, its tide crests shuttle in and out continuously at a frequency that happens to correspond to the lunar tide. This rare synchronization creates a seiche, or standing wave, that resonates back and forth, affecting all travelers, from plankton to ferry-boats.

These colossal forces also foster the extreme speed of the gulf's currents, which run at up to six knots. (The powerful Atlantic Gulf Stream, at its fastest, typically runs at less than four knots.) Speed makes for strange watery phenomena. To the south of Isla Rasa lies the Canal de Salsipuedes, the Get-Out-If-You-Can Canal, a notoriously dangerous collision point of incoming and outgoing water, complete with riffs, whirlpools, white water, haystacks, roostertails, and standing waves, kicking off from deep layers of water and submarine gorges thousands of feet below. The dynamics here are so intense that liquid takes on something of the properties of a solid, with the water on one side of a current standing noticeably taller than on the other side, the line between them forming a ledge as sharp as if cut by a knife.

Of course the dangerous flows of the gulf are dangerous only from our perspective. For full-time residents, they're no more intimidating than the speed of the blood coursing through their own veins. In fact for most marine life these currents *are* their lifeblood. Water transports the denizens of the sea far beyond the capabilities of their own fins and tails—as if our footsteps were augmented or controlled by the wind. Even tiny life forms make enormous voyages in the deep blue home.

On the flip side, because moving water carries so much floating life, some creatures, such as corals, anemones, sponges, oysters, and barnacles, adopt the reverse strategy and don't go anywhere at all, simply anchoring themselves in place to fish on those passing by. In the terrestrial world, only web-weaving spiders fill a similar niche.

For Rasa's birds, the life suspended in the underwater world acts as a continuously replenishing transfusion of lifeblood. Heermann's gulls and elegant terns come to the Gulf of California because of the fish in this life-giving sea, and specifically for the enormous schools of northern anchovies (*Engraulis mordax;* Red List: Not Evaluated) and Pacific sardines (*Sardinops sagax;* Red List: Not Evaluated)—the same species of ten- to fourteen-inch-long silvery fish that fueled Steinbeck's Cannery Row during the Great Depression. Sardine schools, which range as far north as Alaska in shoals millions strong, comprise what may once have been the biggest entities ever to live in the ocean, migrating north in the summer and south in the autumn, each size class roaming farther each year as the fish age over their natural life span of twenty-five years or more.

On Isla Rasa, sardines and anchovies provide the ornithological equivalent of mother's milk, siphoned from the ocean and transferred from parents to baby birds. Each fish is a portable baby bottle: oily, calorific, high in protein, rich with omega-3 fatty acids, loaded with condensed energy ideal for jump-starting a nestling into a fledgling in only thirty-five days (gulls) or twenty-seven days (terns). Eaten whole, bones and all, these fish provide calcium vital for the manufacture of the chick's first home, its egg, and for the assembly of its interior home, its skeleton.

Heermann's gulls feed their young by regurgitating partially digested fish from a saclike pouch in the esophagus known as a crop. Elegant terns carry a single fish crossways in their bills and offer it to their chicks whole. Thousands of tern parents arriving at Rasa's shores at any one time look like flocks of dainty butterflies laboring in flight under the weight of their drooping sardine mustachios.

Rasa's resident gulls and terns and their young eat many tons of fish a day during the nesting season. The consumed fish are then metabolically transformed into nearly as many tons of guano, all or most of which works its way back to sea to nourish the phytoplankton that feed the fish. Thus the birds are fueled by that which they help to grow, the fish are fueled by that which preys upon them, and together they build the energetic bones of the deep blue home.

8

The Unreefed World

LIKE MOST OF RASA'S MIGRATORS, I arrive here at the end of my resources—in my case, with exactly ten dollars in my pocket and only a vague notion of how I'll get home at the end of the field season, or where home is anymore. In my journal, I catalogue a growing list of things I don't now have but will somehow acquire for next season: sunglasses, goggles for the guano dust, a broad-brimmed hat, mosquito netting, a portable awning for shade, a pair of flip-flops (preferably two), hiking boots, baby powder, Cutter bug juice, suntan lotion. I bemoan my species' advancements—the loss of fur, the insistence on bipedalism—which have forced us to invent technological substitutes for all the technology other species carry onboard, no baggage needed.

I have, however, brought a few things worth their weight in gold: a mask, snorkel, and fins. A small umbilicus—an *estero* composed of three tidal lagoons on the northwestern shore—connects the island to the sea. At low tide the *estero* subsides into a puddly sandbox paved with sodden feathers. At high tide it provides the breeding birds a place to rest and relax when not on nest duty. At both times, tired, thirsty, dusty residents refresh themselves here.

The view from our *casita* is dominated by the *estero* even more than by the imposing ridgeline of Isla Ángel de la Guarda or the spine of the more distant and more imposing Baja Peninsula to the west. The *estero* is a combined mirror-sundial-moon-dial-thermometer-barometer-tide-gauge-anemometer, capturing weather, currents, time, lunar phase, bird-breeding calendar, and migratory cycles. We learn to read its water-washed register with intimate precision.

The lagoon is also the island's welcome mat. Thousands of shorebirds use it on migration: sandpipers, curlews, dunlins, knots, dowitchers, godwits, turnstones, surfbirds, tattlers, whimbrels, willets, and yellowlegs. Each employs the lagoon in different ways. Some stitch the sand, their tiny bills the needles. Others pierce long holes with rapiers. Some beachcomb, flipping pebbles, dried seaweed, cactus skeletons, crab shells. The herons, egrets, and night herons freeze in patient poses, some with wings open, extending the deadliest of lures, shade. All these visitors are refueling for long journeys, some flying north to the Canadian High Arctic, some beyond, to the northern shores of Greenland. Part of this island will travel with them.

Whenever the *estero* is full and my time is free, I visit the underwater visitors: needlefish, schools of yellow-striped goatfish, brown sea cucumbers, sea urchins, sea stars, sand anemones, stingrays. Most of these visitors skitter into the *estero* and quickly out again, finding little to their liking in a world dominated by birds.

A quarter of a mile to the northwest of Rasa lies Islote Rasito, a nubbin of rock so whitened by guano it gleams even on moonless nights. Despite its diminutive size, Rasito is one of the loudest places around, a veritable bullhorn in this part of the gulf, broadcasting and amplifying the continuous outraged news from a colony of California sea lions (*Zalophus californianus;* Red List: Least Concern 2008), whose bellows, grumbles, growls, harrumphs, yelps, and howls issue forth day and night, eclipsing at times even the bird sounds.

Immature sea lions from Rasito sometimes venture to Rasa, wiping their fins on the welcome mat of the *estero* before hauling out on the rocks to sun for a spell. They are brazenly unafraid of us. When I swim in the lagoon at high tide, one, lolling on its back, nonchalantly drapes its head across the rocks to study me from upside down, whiskers twitching. Once a young sea lion joins me in the water, its body as lithe as waving kelp, folding and twisting around itself, a contortionist with soulful eyes and mischievous manners. Underwater, it pulls my long hair, making me yelp like a sea lion.

On this day, afloat in the lagoon, I follow a Sally Lightfoot crab (*Grapsus grapsus;* Red List: Not Evaluated) working the rocks at the edge of the pass between the *estero* and the sea. She scales the canyons and rappels the gorges of her world on tiptoe. In pursuit of Ed Ricketts's science, John Steinbeck spent many a comedic hour failing to catch a fleet Sally with a hand net. "Man reacts peculiarly but consistently in his relationship with Sally Lightfoot," he wrote. "His tendency eventually is to scream curses, to hurl himself at them, and to come up foaming with rage and bruised all over his chest."

The Sally at the mouth of the *estero* clambers in toe shoes stained white by guano. She doesn't concern herself with me, only glances my way now and again with lavender eyes unfurling on yellow stalks as she goes about her light-footed work. Far easier for me to float, unarmed, following her rambles . . . to be netted myself.

This afternoon my captor goes overland, through the pass and beyond the *estero* to Rasa's outer shore. She hugs the rocks at water level, where her algae dining table is set most bountifully. I hover in the water close by, mindful of the currents and the great white sharks hunting Rasito's sea lions. The waters run swift and hard, nourishing opaque clouds of phytoplankton that stain the surface waters chlorophyll green. The density of this microscopic life reduces underwater visibility to fifteen feet at best. Beyond that I see only a verdant wall.

But the plankton-eating invertebrates of the Gulf of California have other means of seeing underwater, mostly through taste, and so these thick waters offer no impediment, only sustenance. Such fruitfulness transforms Rasa's flanks into tangled forests of soft corals, the sea fans and sea whips known as gorgonian corals. Unlike the Greek Gorgons for whom they are named—terrible beings with writhing snakes for hair—these are delicate polyps that form lacy colonies of gold, white, and yellow, waving their tentacles into the currents and sipping with receptors as discerning as hummingbird tongues.

The underwater communities of the Midriff are paragons of the art of the soft coral, an entirely different medium from the stony corals that build tropical reefs. Soft corals join a promiscuous catalogue of species that lease the prefabricated architecture of rocky shores and islands. They build nothing, yet their combined energies create a world Steinbeck declared "ferocious with life." These animals of the sea are so different from the animals of the land that we name them for plants or inanimate objects: sea pens, Christmas tree worms, sea fans, cup corals, barrel sponges.

Outside the *estero,* clinging underwater to Rasa's skirts and peering at things on my side of the green wall, I find many cryptic denizens of the Gulf of California's animal-forests. The waters abound with extravagantly colored nudibranchs (Latin *nudus:* bare; Greek *branchos:* gill)—sea slugs lacking shells who wear their external gills flamboyantly: clown nudibranchs flaunting harlequin polka-dotted gills in red and yellow; Mexican dancers wearing flouncy blue flamenco skirts; galactic sea slugs that resemble stellar charts of supernovae. All excel at recycling the defenses of their prey, building chemical "shells" from their meals of poisonous sponges or arming themselves with stinging cells absorbed from nibbled sea anemones.

I dismantle Rasa's rock foundations in search of hidden life, lifting stones and waiting for the puffs of silt to waft away. Other strange creatures emerge: flower urchins wearing camouflage of broken Sally Lightfoot shells, chocolate chip starfish, slipper lobsters. Reassembling the reef, I find a two-inch-tall coral shaped like a tiny baby hand, with three pink fingers of coral joined to a coral palm. Suddenly it stands up and wobbles away, cheerily waving a fuzzy glove of feeding polyps.

Hi, I say through my snorkel.

It's a staghorn hydrocoral (*Janaria mirabilis;* Red List: Not Evaluated), not a true coral at all but a colonial marine hydrozoan more closely related to the Portuguese man-of-war. It makes its living off the community of suspended life carried on the backs of upwellings and currents, capturing zooplankton with specialized stinging cells as effective as fishing nets. The hydrocoral is ambulatory not by its own faculties but by those of its symbiotic partner, the even smaller staghorn hermit crab (*Manucomplanus varians;* Red List: Not Evaluated), who has taken up residence inside the coral's base, at the palm of its hand.

As in all marriages, the two partners profoundly modify each other over time. The hydrocoral chemically erodes the hermit crab's seashell home, dissolving it to nothing. The crab chisels the underside of the hydrocoral into its new "shell." In times of threat, the crab slams the door of its home with the deadbolt of its claw and allows the hydrocoral's stinging cells to fight off enemies. In return, the crab allows the hydrocoral, a sessile (Latin *sessilis:* sitting) being, to become ambulatory, trudging along under its steam and sharing the leftovers from its messy table.

That's not all there is to the marriage, though. Like Atlas holding the celestial globe of the sky on his back, the crab carries a universe, since the hydrocoral is loaded with other symbionts and hitchhikers. A community of the single-celled algae known as zooxanthellae inhabit its tissues, donating to the hydrocoral the byproducts of their photosynthesis — oxygen, carbohydrates — in return for a safe home, a dose of carbon dioxide, and a sprinkling of nutrients.

A tiny colony of barnacles lives on one of the hydrocoral's fingers. Related to crabs, barnacles have abandoned mobility and settled down in miniature six-sided fortresses — which, like the nest scrapes of terns, form perfectly packed hexagonal colonies. Each armored fortress is home to a single barnacle animal, which spends its life lying on its back with its cirri (Latin *cirritus:* filamentous), or feeding legs, curled overhead. At dinnertime the armored doors crank open and the cirri kick like tiny synchronized swimmers, gathering live food from the currents.

The doors also open during barnacle sex, allowing each animal to visit its neighbors without leaving home. Not only are barnacles hermaphrodites, with two sets of genitalia, they also have the longest penis in relation to body size of any animal on Earth, up to eight times their body length. When the mood is right and the waters begin to warm, the tiny creatures snake their appendages out the doors to fertilize their neighbors' eggs. The favor is reciprocated by the same neighbor, by a different one, or by many. Barnacles living alone fertilize themselves.

The wanderings of these hitchhikers through the animal-forests of Rasa's outer shore are dictated by the crab. The benefits of attaching oneself to the peregrinations of another outweigh the risks, apparently, since the number of biocoenoses (Greek *bios:* life; *koinos:* common; meaning *community of life forms*) living in, on, under, and around hermit crabs is enormous. At present count, and excluding parasites and flora, more than 550 invertebrates from 16 phyla associate with more than 180 species of hermit crabs.

For those born to the sedentary lifestyle, anchored in place, waiting for drifters, it pays to affix to a wanderer, to reap the benefits of mobility while paying only the costs of immobility. The flow of *rasa* travels thrifty paths.

9

The Epitome of Unrestrained Freedom

W E'RE SEATED ON CRATES in the tiny kitchen of the *casita*, drinking coffee, eating leftovers from dinner for breakfast, writing in our journals. The wind is blowing hard, and the Ramones are playing on the tape deck loudly enough that we don't hear the approach of footsteps. A man's head appears in the doorway—drooping mustache, dark hair, mischievous eyes. He looks surprised to find us here. We're surprised to have a visitor. We rise, kicking over crates, journals falling to the floor. Behind him is a long line of people bearing the trademarks of American tourists: outsized sun hats, sun shirts, hiking boots, binoculars.

It's Tim Means, Californian by birth, Mexican by choice, friend to researchers in these parts, and already a legend in the gulf in 1980 as the founder of one of the earliest ecotourism outfits anywhere, Baja Expeditions. His guests would like to see the tern colony, he tells us. We're happy to act as guides, mostly because we want to make sure they stick to our paths.

The guests are chatty and clean, fresh with the nearly forgotten scents of shampoo and laundry soap (we wash in the sea), and full of questions, mostly about how we live. We herd them gently across the island, cinching them close with our stories, explaining the nontechnological wonders of our outhouse, our rat-prevention measures, the grids we've marked in some of the valleys, our system of paths. We offer mimeographed handouts on bird-viewing protocol and intersperse our directives with mobile lectures on the island's geology, flora, history, and ecology. Tim is bemused and silent in the face of our tour-guiding. He is the most laconic of men, speaking in silences or, when he's feeling talkative, with a lift of the head or a jut of the chin.

He is also accustomed to the responses of his nature-loving guests, whose affection for the outdoors visibly evaporates in the heat and dust of the tern colony, soon replaced by amazement, then horror, then a jumpy irritability as they fan themselves with our handouts and stare over their shoulders toward escape. Given enough time, most guests turn their backs altogether. It's too much: too loud, too hot, too smelly, and the *bobito* gnats are as thick as smoke. They want only to return to the pleasures of the *Don José,* Tim's brand-new live-aboard nature-viewing boat. Tim, however, does not believe in pampering. This is the wilderness, that thing his guests signed up for. They must endure it a while longer and maybe learn to love it for what it really is. He settles onto my favorite flattish rock, elbows on knees, squinting into the glare, chin jutting silent volumes.

To be fair, not all nature-loving visitors react this way. But those who do prove beneficial to the three of us residing on Isla Rasa. Compelled by pity, they invite us aboard the *Don José* for lunch or dinner or both, offering showers, fresh food, cold beer, and valuable portions of their secret stashes of American junk food. Their presumption of our misery results in our returning home to Rasa fatter, drunker, cleaner, and more committed to an island that feels quieter in the absence of human chatter. We come to see the arrival of guests on Isla Rasa as akin to a beneficent current or a periodic high tide carrying nutrient bonanzas our way.

Tim proves our greatest benefactor, supporting us in many quiet ways, not least by rearranging the *Don José*'s schedule to accommodate our visits to neighboring islands. Though anchored to Rasa, we are delighted to temporarily affix ourselves to his boat in order to broaden our observations of gulls and terns hitchhiking on the winds in pursuit of prey suspended on the currents.

He invites us north to the two-part island known as Isla Partida, our closest neighbor on an island-studded horizon. One half of Partida is a double-humped monolith of rock whitened around the edges with guano, and the other, a sloping low island sometimes known as Isla Cardonosa. Compared to Rasa, these are quiet islands, with no vibrating, jittery clouds of birds—though when we drift closer, we're greeted by the familiar tremolo yelps, *yeow*s, and *hahaha*s of yellow-footed gulls.

Isla Partida harbors two wondrous secrets. The double-humped island is a jumble of talus slopes: enormous rock piles formed by the chaotic accumulation of softball-sized stones eroded from the steep hillsides above. The talus appears utterly barren. But when we climb, it awakens and begins to sing—squeaky, chattering, whirring songs interspersed with wheezy breaths. Tim listens, chooses a spot, sits, and begins to lift the stones of the scree, gently, one after another. A few layers down, we see what he's after, a dusky black bird only five inches long, nestled in the protective womb of rock. He reaches in, cradles the unprotesting creature in the palm of his hand, and lifts it out so that we can discover its secret: a single oval egg the size of an almond, pure and pearly white in this dark place where no camouflage is needed.

The bird is a least storm-petrel (*Oceanodroma microsoma;* Red

List: Least Concern 2008), the smallest species among the smallest of all the seabirds, members of the family Hydrobatidae (Greek *hydor:* water; *bates:* treads; meaning *those who walk on water*). Few landlubbers have ever seen a storm-petrel, though the deep blue home supports twenty-plus species. Offshore they're as familiar as sparrows—dark birds, most with white rumps, flitting close to the surface, wings open, legs pattering, stirring with feet and dipping their heads to gather small prey. No other birds at sea act like them. From a distance, they bear the fluttery signature of butterflies drawn to the nectar of the surface.

Within the Hydrobatidae, however, there are recognizable differences. Some species hover on rapidly beating wings. Others dynamically soar, gliding across wave fronts to gain lift from the vertical wind gradient. Others slope-soar, turning into the wind to rise, then gliding back to sea. The white-faced storm-petrel pogos, holding its wings motionless into the wind and pushing off the water in a succession of bounding jumps.

Some of these endlessly roaming seafarers are drawn to boats, appearing alongside in loose, bouncy flocks. They've acquired many nautical nicknames: "petrel" is a diminutive for Saint Peter, who, legend says, walked on water; "storm" refers to their habit of sheltering from foul weather in the lee of vessels. Seamen of old called them Mother Carey's chickens—a derivative of *Mater Cara,* a name for the Virgin Mary—because they believed these tiny birds to be the souls of drowned sailors, whose presence foretold storms.

Griffing Bancroft visited Isla Partida in the 1920s and described least storm-petrels as the most gentle and harmless of all water birds. "Timid and diffident and self-effacing they scarcely struggle if handled, and yet, when harrying the ocean far from sight of land, they express the epitome of unrestrained freedom." He and his crew dug waist-deep trenches in the talus in search of eggs.

In the aggregate our party . . . took nineteen. We did so through the exposure of a many storied tenement house in which floors consisted of a few square inches of weathered sand and doorways of narrow crevices, devious, perhaps, but not quite closed . . . There was acre upon acre of talus, congested with these . . . forms of life to greater depths than we could reach, an impregnable citadel that will ensure the perpetuation of the inhabitants for a future longer than mankind can see.

In the 1920s Isla Partida, surrounded by a deserted sea, appeared invincible. But in 1980 we know that only three islands in all the world host nesting least storm-petrels and that Isla Partida is home to nearly 95 percent of the species, making this island no citadel but a precarious life raft of survivors.

Our timid and self-effacing bird sits quietly as Tim hands her over to Enriqueta, who opens one wing, then another, extends one leg, than another, looking for bands—the mark of someone else's scientific endeavor. She blows onto the chest, parting the outer feathers to reveal the brood patch, the featherless, blood-rich area that develops in most birds a few days or weeks before egg laying. The cherry-red skin warms first the eggs, then the chicks, a swaddling blanket with a heartbeat.

Nearly six decades have passed since Bancroft dug for least storm-petrels here. We do not do with our captive what he did with his when he'd finished his examination, that is, toss it into the air. "The Duck Hawk [peregrine falcon] overhauled this one as though it were standing still, as though it were just fluttering and making no effort to escape," wrote Bancroft, in genuine surprised sadness that a butterfly-of-the-sea could not outmaneuver a falcon. Nor could he imagine how any species at home on these desert islands might fail to outmaneuver the human footprint bearing down upon them.

Enriqueta places the bird gently on its egg. Tim rebuilds the rock ceilings and walls. We tiptoe away. A few years from now, the growing etiquette of ecotourism will prohibit digging through the talus, granting the birds the privacy they seek . . . yet denying the coming generations of humans another connection with nature, another lifeline between us and the wild others. Maybe the one that will save us all.

10

Mirage

LEAST STORM-PETRELS RETURN to breed on the same island, and likely even in the same rock pile, where they were hatched, a lifestyle known as philopatry (Greek *philos:* beloved; Latin *patria:* fatherland). They spend their nonbreeding lives entirely at sea, either in the air or on the waves, where dangers certainly abound: large fish snatch them from the surface; frigatebirds, falcons, and ospreys hunt them from the air. Yet no offshore risk compares to the risks onshore, since these tiny birds are effectively crippled on land, their legs set so far back on their bodies that they are forced to crawl on their ankles and the crutches of their wings. Once ashore, they fall victim to all hunters, from cats to rats to dogs to yellow-footed gulls, who swallow them whole—little feathery *amuse-bouches.*

Storm-petrels avert these calamities by adopting a clandestine lifestyle—nesting underground in burrows, caves, or talus and going to and from these safe houses only under cover of night, when gulls and their like are asleep or at least not hunting. Each parent takes turns fasting and incubating their single egg in shifts lasting up to six days, while the off-duty parent hunts at sea, replenishing itself. The nest relief, or trading of places, occurs at night in a ceremony accompanied by much twittering, chirping, squealing, and cooing, though before that can happen, the returning parent must find its nest in the enormous colonies favored by storm-petrels—in the case of Leach's storm-petrels, off Newfoundland, in a colony housing more than 7 million birds. Returning birds fly up and down the slopes, calling and listening for the calls of their mates underground. Once reunited, the two perform a purring duet that sounds remarkably like sighing—which, perhaps, it is. All the storm-petrel calls put together make a strange, unearthly, and unforgettable nighttime soundtrack.

Despite their diminutive size, storm-petrels are among the most industrious of avian parents, enduring lengthy incubation periods of up to fifty days, after which the newly hatched chick is brooded continuously for a week or more before being left alone in the nest during the day. Thereafter, both adults fish all day and return to the nest at night, each laboring to bring home a meal weighing up to 20 percent of its own body weight, comprising a rich trail mix of the sea, complete with small fish, squid, zooplankton, and a musky stomach oil as calorically valuable as milk. The oil, formed from partially digested prey, is nutritious enough that each parent needs to feed its chick only once every twenty-four hours. The daily meals are repeated for fifty to seventy days, until fledging.

The storm-petrels' enormous shared parental investment seems to strengthen the pair bond. Despite their long lives—some of these little birds clock life spans of thirty-six years—auditioning a new mate appears to be more work than it's worth on their tight time-and-energy budgets. Most, if not all, of the Hydrobatidae are monogamous, and their young are among the few who consistently pass the DNA paternity tests failed by so many other supposedly faithful species, including our own.

· · ·

Ecologists call storm-petrels K-strategists, a term derived from the r/K selection theory, which posits that environmental pressures foster one of two survival strategies. The K-strategists live many years and produce few offspring, which require much parental care. The r-strategists have short lives and produce many offspring requiring little or no parental care (think locusts). K-strategists arise from stable, predictable environments, r-strategists from unpredictable or ephemeral environments.

In general, K-strategists tend to be larger animals, such as humans, elephants, and whales, though diminutive storm-petrels succeed with this strategy too, thanks in part to their proclivity for nesting on islands. Conditions on Isla Partida have likely changed little in thousands of years, providing the kind of constant that K-strategists require to fare well.

Change, however, derails them. The Guadalupe storm-petrel (*Oceanodroma macrodactyla*; Red List: Critically Endangered 2008) nests only on remote Isla Guadalupe, 150 miles west of the Baja Peninsula. For millennia this island truly was an impregnable citadel, 92 square miles of unique species and habitats, including forests of Guadalupe palms, Guadalupe cypresses, Guadalupe oaks, California juniper, and Guadalupe pines. The Guadalupe storm-petrel nested in burrows under rocks and between the roots of the pines in cloud forests above 3,000 feet. No terrestrial predators shared their world.

That is, no one did until Russian whalers and sealers arrived in the nineteenth century, bringing with them Earth's most ruthless hunters of plants and animals, domesticated goats and cats. By 1906 the visiting ornithologist W. W. Brown, Jr., found a nearly unrecognizable landscape. "The whole island seems threatened in the near future with absolute desolation—doomed to become a barren rock. The mortality among these birds from the depredations of the cats that overrun the island is appalling—wings and feathers lie scattered in every direction around the burrows along the top of the pine ridge."[1]

The storm-petrels, who bid their all on a nocturnal, underground way of life, proved no match for land predators with nighttime talents. Brown noted that the birds were still breeding in large numbers on Guadalupe and that "sometimes at night the air seemed to be fairly alive with petrels, their peculiar cries being heard on all sides." But not for long. Those birds who escaped the cats fell victim to the secondary predations of the feral goats, whose population, fueled by virgin vegetation, ballooned to 100,000 before crashing to near nothing as they ate the island to the ground. Goats chewed the California juniper into extinction and the pine trees in the cloud forests to near extinction. Without the tree roots to hold them in place, the storm-petrels' burrows collapsed and washed away.

In 1928 Guadalupe was designated one of Mexico's first nature conservancies. But the goats and cats remained. In 1989 visiting researchers found packs of extremely aggressive feral dogs roaming the island, adding to the woes of the surviving birds. By 1994 the Red List catalogued the Guadalupe storm-petrel as critically endangered, possibly extinct. The worst seemed confirmed in June 2000, when a team of scientists from the San Diego Natural History Museum visited the island during what should have been the bird's breeding season, found none, and concluded that the species is indeed extinct.

In the decades since the whalers and sealers and their symbionts arrived, five other bird species on Guadalupe have succumbed, along with at least four kinds of plants and along with habitats in their entirety, including the California juniper woodlands once endemic to the island.

The swing from health to extinction can be brutally rapid. Ella Vázquez-Domínguez of the Universidad Nacional Autónoma de México's Institute of Ecology found perhaps the most striking known example of the ephemeral nature of life. In 1995 Isla Estanque, a tiny island twenty-five miles northwest of Isla Rasa, supported a healthy population of an endemic deer mouse (*Peromyscus guardia estanque*; Red List: Critically Endangered 2008). Three years later a feral house cat was spotted on the island. Its scats were collected and analyzed and found to contain the hair and bones of deer mice, and one year later the cat was eradicated—too late, apparently, for the deer mice, who have never again been seen on Isla Estanque. Vázquez-Domínguez believes the entire population may have fallen victim to a single introduced cat.

Islands are invulnerable until they're breached. And then they're as fragile as mirages.

11

Emotional Ecology

IN THE UPANISHADS, the classical Hindu texts on medita-
tion and philosophy, *rasa* has yet another meaning. It refers to
the essence or flavor of something with great emotional effect,
something existing beyond the senses. Classical Indian art as-
pires to arouse nine *rasas*, or emotional flavors: *shringaar* (sen-
suality), *raudra* (anger), *hasya* (comedy), *vibhatsaya* (disgust),
veera (heroism), *karuna* (compassion), *bhayanak* (fear), *adab-
huta* (wonder), and *shanta* (peace). To bring *rasa* to fruition, observers are needed—those who listen to the music and poetry,
watch the plays and dances, observe the sculptures and paintings. The audience participating in the juicy emotions is known
as the *rasika*—without whom there can be no *rasa,* since even the
most skilled artist is limited by the depth of the audience's feeling. Thus art exercises the palate of the audience, whose develop-
ing response intensifies the flavor of the art.

In much the same way, our life on Isla Rasa primes us for
the next act to play out on the desiccated talus of Isla Partida.
The island's second secret wonder is the thousands upon thou-

sands of Mexican fish-eating bats (*Myotis vivesi;* Red List: Vulnerable 2008) inhabiting the world under the stones. They live only in the Gulf of California and on the Baja Peninsula, roosting in rock crevices and beneath flat beach rocks on shores free of terrestrial predators. On Isla Partida they scurry past the storm-petrels, crawling on webbed fingers and big feet, into the infinite network of miniature tunnels threading through the scree. Most of the bats sheltering here in this season are females, who spend the daylight hours nursing newborn pups wrapped inside the cloak of their wings.

Mexican fish-eating bats are the largest of the New World species of *Myotis* (Latin: mouse-ear), with wingspans up to fifteen inches. In flight terms, they possess a high aspect ratio, meaning their body is wider than tall, and a low wing load, meaning a large wing area relative to weight. This enables them to fly low and slow while carrying large payloads in the form of fish. These are not the bats of your backyard world, fluttering with impossibly aerobatic maneuvers in pursuit of insects. Mexican fish-eating bats lumber with the slow maneuverability of box kites.

Rather than nimbleness, they possess stamina and strength. *Myotis vivesi* fly many miles a night, far out at sea. They visit Rasa's *estero,* along with all the other lagoons, bays, and inlets in the Midriff Islands region. When weather and waves permit, they patrol the open waters of the larger gulf, echolocating for telltale ripples on the surface and snaring small fish and invertebrates with large gafflike claws on their hind feet, then flipping their prey into their mouths to be eaten on the wing or carried to shore. It's a life unlike that of any other mammal: sleeping on land, traveling on the air, feeding from the ocean—the stuff of dreams for many an Earth-locked human.

Enriqueta and I return to Isla Partida at nightfall. A few of the *Don José*'s guests join us, riding a *panga* to shore as a full moon rises above the distant ridgeline of Isla Tiburón. The wind is blowing hard from the south, perhaps the first bluster of the impending North American monsoon: summer rainmaker for Arizona and New Mexico, triggered when sea-surface temperatures in the Gulf of California reach 79 degrees Fahrenheit. The winds are strong enough that we wonder if the bats might be grounded, since big wings are unstable in even moderate blows, and these winds are strong enough to drive breaking waves onto the beach, wetting us all.

Our landing is a through-the-rabbit-hole gateway. We step into the surf, dragging our feet through the sucking gravel, look up, and there they are. The air is a black blur of wings. There is no component other than wings, thousands upon thousands stirring the wind so that the air breathes from many directions at once. The flyers are as thick as locusts, noisy with whirring stridulations, wings flickering strobelike in the moonlight so that everything appears jumpy and in slow motion—even though the action unfolds faster than our senses can deconstruct it.

One thing we can see is the difference between bats and storm-petrels. In the moonlight, the bats climb out of the talus and unfurl big skin sails, rising like parasailers before cutting the tow rope with the land and echolocating past us on clicks as bright as submarine *pings*. The petrels climb out of the talus and flop, as A. W. Anthony wrote in 1900, something like an old felt hat before the wind.[1] Crippled on the ground, they loose eerie cries, strange, whirring *krrri*s and bansheelike *tuc-tuc-a-roo*s—the wailing atmospherics of a distant radio station being tuned. Some birds shriek before they jump up to launch into flight, which in an instant transforms them, making them more batlike than the bats—aerialists with black cloaks that conjure magic tricks of agility and speed. The storm-petrels change tack with each deep wing beat, zigzagging over the water, zippering the darkness closed behind them.

The sounds and sights, the scents of birds and mammals, the dampness of the south air, the oscillating winds, the moonlight, infect us with some kind of craziness. Enriqueta and I race to the top of the island, leaving the guests on the beach. We stumble, crawl, crab, and climb our way over rocks, up boulders, past cactuses, around cliffs, along the dangerous edge where not even the birds and bats will nest. We move inside a bubble of black wings and deafening shrieks, pushed by the wind. We climb into an intoxication of birds and bats, each with their own persuasive, infectious life force.

Our ancestors mimicked the talents of other species, learning to dance and sing like them, to draw and paint them, to tell stories of them, to decorate themselves like them—the first expressions of the emulations we now call art. This night, climbing hard and fast, Enriqueta and I reach the island's upper slopes, where cárdon cactuses stand like giant ghosts and from where we see the *Don José* far below, lights twinkling. Birds and bats come with us. Breathless, uplifted, we float on their black stars, offering the *rasika* of our bodies—our audience for their life's performance.

12

The Anti-Bodies of Quiet

IN HIS TRAVELS THROUGH THE GULF, Steinbeck wrote of how quickly the big world dropped away here. "The matters of great importance we had left were not important. There must be an infective quality in these things. We had lost the virus, or it had been eaten by the anti-bodies of quiet."

On Isla Rasa, the dramas of the great world are replaced by more intimate stories. By the end of May, the courtship flights, matings, and incubations have culminated in a landscape overflowing with new life. The gulls raise bountiful broods of two or three chicks, who surreptitiously expand their parents' hard-fought boundaries by sneaking into the shade of nearby cairns or rock walls, where they pant for relief from the summer heat. They arouse in us the *rasa* of *karuna,* compassion. During the cooler mornings, the chicks prepare for flight, hopping up and down, flapping half-feathered wings, experimenting with lift, transforming the entire island into a stage of jumping-bean players. They arouse in us the *rasa* of *hasya,* comedy.

The terns experience a different childhood. Within a few days of hatching, each chick is lured from its nest by a parent performing the fish-offering display—a ruse designed to introduce the fuzzy youngster to one of the roving bands of chicks known as crèches (Old French: crib). Crèches are seething, tightly packed masses of baby terns, tended by only a few adults, freeing the remaining parents to forage offshore.

The tricky part, from our *rasika* viewpoint, is the business of a parent tern locating a single roaming chick in a mass of seemingly identical roaming chicks. At any given moment, birds by the hundreds are circling the crèches, each with a fish in its bill. Although they have likely flown many miles to and from their foraging grounds, they are logging even more miles now. The parents trill loudly, the chicks call back, and the decibel level in the ternery rises beyond comprehension. When contact is finally made with the right chick, the parent drops into the center of the crèche to deliver its one perfect sardine or anchovy.

Watching from my flattish rock, I feel the boldness of this accomplishment and of all the cumulative accomplishments that have led both gulls and terns to this point in a breeding season spiraling toward completion on Isla Rasa.

The antibodies of quiet enliven our personal dramas too. Returning to the island from our night out aboard the *Don José*, we find Pinguino, our supply-boat driver, waiting for us. He has been dispatched by Antero Díaz, our worried benefactor in Bahía de los Angeles. Caught up with birds and bats and the social whirl of a boatload of twenty visitors and eight crew, we have missed our usual six o'clock radio transmission two nights running.

The *panguero* is happy to see us alive and well, and when we invite him to the *casita* for dinner, he blushingly accepts—a rarity. We dig through the resupply, hungry for fresh tortillas, mail, and whatever treat has been included for the night's meal, usually fresh fish. To our surprise, we find a gift from Don Antero himself. He is the most prominent man in the most prominent family in this *minúsculo* village, erstwhile mayor, postmaster, police chief, the larger-than-life founder of Casa Díaz, the first hotel in Bahía de los Angeles. He has been a friend to scientists in this region for many years. Now, especially for us, he has sent along a thick steak of *borrego*, the bighorn sheep (*Ovis canadensis cremnobate*; Red List: Least Concern 2008), with a note saying he hopes we enjoy this wild meat of the mountains.

We stare at the bloody wrapping, caught in a small swirl of *vibhatsaya*. The hunting of *borrego* is outlawed. Everyone knows this. Antero also knows we are most likely *ecologistas* as well as scientists. Confused, discomfited, we discuss his motives, deciding they are a mixture of challenge, friendship, generosity, dissent. A trickster test. We debate what to do with the meat, factoring in the probability that Pinguino accepted our dinner invitation in expectation of the fresh game. In the end we decide there's no point wasting the food, so we cook it and serve it, or rather Mónica, the only chef among us, cooks it and serves it with curried mashed potatoes and mushroom gravy.

The list of *rasas* does not include guilt, the secret ingredient tainting my meal. In my journal I record that the food is uncomfortably heavy, underlining the word *heavy*.

Questions of human impact, whether from *borrego* poachers, egg collectors, or ourselves, begin to insert themselves into our work. How much should we disturb this world, even in the quest for knowledge? The three of us are not in agreement. Initially, our disparity amounts to little more than a reexamination of what we thought we knew of each other. Eventually, philosophical differences pry our party of three into two camps, a split as painful as an amputation.

The crack appears early, in the first nights after raising the radio antenna, when we realize that gulls occasionally fly into the guy wires and break wings and necks. Unable to sleep to that unhappy soundtrack, Enriqueta and I resolve to paint the lines a fluorescent yellow. It's not much, but it's all we can think of. Mónica disagrees with our concern. This is not the first time science has caused the death of study animals, she says.

Many small points of friction come to a head in the unbearable heat of June, when Mónica declares her intent to enter the tern colony and band the chicks. She also wants to build a corral to surround an entire crèche, so she can band the chicks and recapture them for long-term data collection.

Because I'm working with Mónica, she expects me to help. Yet I doubt the wisdom of these plans. The operative component of a crèche is its mobility. Contain it, and the gulls and ravens and falcons will pick off the flightless youngsters with ruthless efficiency. The incursions from the lobster fishermen and the serious loss of tern nests this season lead me to believe that we should not attempt any invasive research. Mónica disagrees. We fight. We cry. Come morning she tells me I'm fired and should leave on the next *panga*. Enriqueta, peacemaker, offers me the opportunity to stay and work for her. Mónica vows to band the terns and build the corral anyway.

The *casita* is too small for the three of us now, and Mónica moves into a blind on the edge of the tern colony. The monsoon arrives for real, roaring upon the island like wet fire. It's too hot to move during most of the daylight hours and nearly too hot to breathe. It's so wet we pray the skies will shed rain. But they never do. Their bounty is gripped tight all the way to landfall on the mountaintops of Arizona.

Under Enriqueta's tutelage, my notes accumulate heft and sweat. We work at dawn, dusk, and after dark—though much of our investigation requires us to make midday counts or observations too, so we suffer those hours as well. We're slowed nearly to a crawl by the oppressive breath of the desert and the weight of the sea air. Perpetually exhausted, we guzzle electrolyte-rich Pedialyte straight from the bottle and order as many more bottles as Bahía de los Angeles can muster.

A handful of Heermann's gull chicks move into the shade of the *casita,* jumping and flapping at our feet like a junior kickboxing class. I give them the dried Mexican beef called *machaca* and water. The chicks don't know what to do with the mercurylike water and stare, first with one eye, then the other, before puncturing it with their bills. I name each of them in turn Jésus and María.

A new miniature colony of terns begins to form at the northern end of the easternmost valley. A few terns displaced by the egg poachers are trying to lay eggs again. Or subadults are trying for the first time. It's too late in the season, of course, and the effort will come to nothing. But it tells us something of the birds' resilience. They squat in the heat, wings splayed. They pant so fast their feathers oscillate. Their endeavor blurs into a mirage of hopeful misery.

13

Everything Is Already Brilliant

O**N A SUNDAY IN LATE JUNE** Enriqueta and I make a final circumambulation of the island, checking Rasa's other nesters. The peregrine falcon chick, long since fledged and now bigger than her father—therefore likely a female—spends countless hours aloft, studying the language of the air. Already she's fluent at static soaring, holding motionless in the updrafts kicking off from the obstructions of Rasa's cliffs. She flouts the crushing midday heat by parsing thermals, disentangling the currents from the winds to spiral upward beside her parents into thinner, cooler air. Up high, the three chitter excitedly, *chi-chi-chi-chi-chi* calls that fall to Earth like broken glass. Their shadows rake the island, stampeding the nesters below.

We know something of the fledgling's feeling of achievement. If we wanted to, we could traverse the island in a quarter of the time it took us when we first arrived, on the sure-footed billygoat legs we've developed during our months in residence. Enriqueta's blue hat, now bleached by sun and countless guano hits, advances through the landscape like a bobber on an invisible fishing line, dipping as she hops between rocks, circling as she scales boulders. Every few minutes she pulls up short, caught on a hook of interest. The truth is, with all we now know and all we want to investigate further, we need four times as long to circumnavigate the island, even with our tougher bodies.

Our eyes have grown stronger too. We raise our binoculars to glass subtle clues at a great distance—in this case, three unusual birds resting on the water far offshore. Could it be? We page through our dog-eared field guides cum wilderness cushions, punctured by cholla quills and volcanic rocks, to the section on petrels and albatrosses. Sure enough, there they are, the gull look-alikes known as northern fulmars (*Fulmarus glacialis;* Red List: Least Concern 2008)—rare avian visitors from Arctic breeding grounds. These three are bathing vigorously, setting up a splash field as good as a waterfall, which snags the curiosity of a few Heermann's gulls, who paddle over to investigate, as well as a half-dozen terns, who flick their wrists to veer from their flight paths and hover overhead, surveying the potentialities.

Between us and the fulmars, in the shallow waters of the *estero*, a reddish egret runs, jumps, and spins through knee-deep water strewn with feathers, flashing its wings, red crest flipping. It's a mad Isadora Duncan dance in the middle of nowhere, entrancing us, its spellbound *rasika*. In the midst of its performance, a prehistoric head punches through the water just offshore. The green blindfold of the surface is ripped away, and a sea turtle rises, the bubbles of its exhale boiling up, followed by twin nostrils, a parrotlike beak, the obsidian pools of its eyes. As it rises to the surface, the egret steps into the air, paddles its wings, and rows away, perhaps to another *estero* on another island.

We study the turtle through binoculars. It's a loggerhead (*Caretta caretta;* Red List: Endangered 1996), a giant from the deep—though this individual, at only two and a half feet long, is still a youngster. Already it's endowed with crushing jaws in a heavy head and already its petite dimensions are amplified by the multispecies crew in attendance: three striped remoras, a constellation of warty barnacles, a tiny school of juvenile fish, and fields of algae staining green the groutlike seams between the scutes of its carapace. There is much else we can't see, since loggerheads provide a movable habitat for more than a hundred kinds of animals from thirteen phyla, along with thirty-seven species of algae.

Terns flutter past, some carrying fish. The turtle's eyes, as dark as the abyss, never waver, staring directly at us, quizzical and otherworldly.

And then it's gone, head retracted into the shell of the ocean, front flippers rowing below the surface. I watch as long as I can: the dark shadow sculls Rasa's flanks, pausing here and there to nibble the bottom for clams and crabs, delicately grinding with keratinous (Greek *keras:* horn) beak, turning the shelled cornucopia of the shore into turtle power.

Because we're moving slowly, we have time to discover that the breezes riffling the water to the south are not breezes at all but a huge school of small fish, presumably *sardinas.* California sea lions, with their distinctive shiny fur and porpoising leaps, are closing in from the west. Behind them blossom the splashes of an enormous school of common dolphins (*Delphinus delphis;* Red List: Least Concern 2008).

As it often does, the action converges not far from Rasa's shore and crescendoes into a full-blown feeding frenzy. Shark fins slice the air. Mahi-mahis (*Coryphaena hippurus;* Red List: Not Evaluated), whom the Mexicans call *dorados,* power through at speeds of 50 knots. Three 70-foot fin whales (*Balaenoptera physalus;* Red List: Endangered 2008) join the fray, the huge mountains of their backs appearing and disappearing like rolling backdrops of glinting scenery. Knots of silver sardines braid and unbraid themselves. Rafts of gulls launch off the island to pluck at the surface.

The gulls congregate around jerky lines of brown pelicans arriving in single file from Isla Salsipuedes, their breeding island to the south. The pelicans crank upward one by one, tilt over one by one, and dive, shoulders drawn, wrists bent, legs tucked. A second before they hit the water they twirl to the left, protecting the trachea and esophagus built into the right side of their neck.

A pair of stiff wings beats toward us, pumping hard, big feet dangling with a heavy load—an osprey carrying a pompano in its oversized talons. The bird sails its catch a few feet above our heads—so close we can hear the fish gasping for air and can stare into its eyes. The osprey flares her wings and drops onto a nearby cairn. There is no coup de grâce in her repertoire. She proceeds to eat the fish alive, starting with its lips, then the unblinking silver eyes. When she's consumed half of it, she lifts her tail, defecates, flaps her wings, and rises with the remains.

We complete our circuit of the island and note the final verdict for its yellow-footed gulls. Despite the ceaseless efforts of the parents and the seeming bounty all around, their chicks suffered 100 percent mortality. Enriqueta already suspects the reason for this failure, and a few years from now she and the biologist Jesus Ramírez will spend a season eradicating Rasa's rat population—which probably arrived with the guano miners of the last century. In the aftermath of their effort, the number of nesting seabirds will blossom to numbers not seen here in more than a hundred years: the 30,000 elegant terns of 1980 surpassed by 200,000 today—as the birds who chose this island for its safety finally regain it.

The swirl of *rasa,* in all its many meanings, flows on. It removes the chicks who succumb to the appetites of the predators and delivers the insects lost on monsoon winds. It powers the moon and the tides.

It carries piles of limp passions, wrung out and used up. We fold and unfold them, looking for quiet. Mónica does not appear to be banding the terns or building a corral after all. Alone, hidden in her blind at the tern colony, she is surrounded by the flow of white chicks motoring fast across hot ground.

We three sleep now in separate quadrants of the island. The nights are so hot that I abandon my hammock for the beach. We are beginning to imagine the future away from here, and much of our imagining is done alone in the dark, filling our respective airspaces with sweat and worry. I doodle in my journal, sketching Enriqueta on her cot, me on the beach, Mónica curled around the rocks in her blind, bubble-thoughts arising from our heads, filled with unfinished mathematical equations and peppered with question marks.

And then it's over. The birds are streaming out to sea to fish and not coming back. The island empties, hour by hour, until only the orphan chicks are left, flapping their wings in our shade. The island strips down to its desolate summer nude, freckled with the carcasses of the dead.

Two *pangas* come for us. Two drivers, but not Pinguino. Perhaps he thinks he will miss us too much to say goodbye? We motor away on a high tide on the summer solstice. Behind us the jittery skies of Isla Rasa fade into the heat. In the end, the three of us depart together, having made a small *shanta*, peace, in the course of packing. We all plan to return to the island for next year's field season. Only Enriqueta will.

In the Canal de las Ballenas, the Channel of the Whales, a line of firecrackers appears on the horizon to the south. The day is without wind or breath, and the sea is oily with calm, soaking up the huge red scenery of Isla Ángel de la Guarda on our right, the purple mountains of the Baja Peninsula ahead, craggy clouds far to the west over the Pacific. The firecrackers flash with the telltale black sparkle of marine mammals—whether skin or fur, we're not sure. We know only that there is a huge number, shattering an enormous glassy sea.

The *pangas,* laden with our gear, lumber westward, while the splashes drive north—firecrackers arching through the sky, somersaulting, flipping, back-flopping, belly-flopping. Our paths are destined to intersect. The splashes materialize into a school of common dolphins, gunshot breaths peppering the air, sleek bodies casting through the surface, the hourglass patterns of their black-and-white-and-yellow skin weaving a mesmerizing geometry underwater.

Their numbers are staggering. For months I've been counting big congregations of wild animals, but now I can't keep up. Tens of thousands? Hundreds of thousands? In the Mediterranean, common dolphins form superpods 300,000 strong. Off New Zealand, superpods have been reported spanning nearly 500 square miles of sea. This gathering in the Canal de las Ballenas stretches as far as we can see, churning the water and air, sleek bodies soaring, tucking, nosing underwater. Mothers with tiny babies jump in unison. Whole lines emerge and disappear simultaneously. Thousands of bodies surround us and refresh us with their spray. Countless swimmers slip alongside the *pangas,* whistling, rolling, and staring through the surface at us, first one eye, then the other. A few ride our bow waves all the way into Bahía de los Angeles.

When we finally step ashore on the North American continent, we are time travelers brought forward in the blink of a three-hour boat ride to a modern world made modern only in relation to Isla Rasa. Even by the island's raucous standards, the sleepy fishing village seems noisy, because its sounds are those we have forgotten: squealing children at play in the bay, a pickup truck rumbling down sand streets, a scolding woman, four giggling *pescadores* mending their nets on the beach, cattle lowing. The sounds of generators and cooking stoves and fans and outboards and hammers and dogs and roosters and babies and slamming doors and pouring water and a guitar step on each other and abut each other in a confusing orchestra.

We reel from the heat blasting out of the peninsula's mountains. It's hotter and drier than on Rasa, hot enough to make us gasp. We raise our binoculars to the mountains, the sea, the sky, looking for distant relief. Of the many adjustments we face upon our return to the tropic of civilization, the most difficult for me will be my inability, for reasons of propriety, to constantly call the world into closer view through binoculars. For a long time everything will seem out of focus and too far away. Likewise, I will miss the stabilizing effect of hours spent advancing at a snail's pace, examining, measuring, taking notes, snapping photographs, speculating.

For one hundred days I lived anchored in the flow of *rasa*, counterbalancing two decades of modern life sheltered from winds, waters, the night sky, silence. The wilderness revived my native tongue: a protolanguage, words without words, floating into perception, submerging. I learned to *be* them: patience, surprise, contentment, foreboding, the natural, the supernatural, suspense, relief, fear, peace. My eyes saw verbs in motion. My ears heard the percussion of nouns. Scents became sentences. My mental treadmill, sprinting and slumping in exhaustion since birth, slowed. Apparently I didn't need to chase after brilliance. Everything was already brilliant.

This was more than a transient sensation. Surrounded by seabirds and the sea, I grew suppler, more grateful, afloat on life's buoyant preserver.

There is yet another Sanskrit meaning for *rasa:* delight in existence. On Isla Rasa, living deeply in the flow, I found myself delighted at my first homecoming to the planet where I was born.

PART TWO

~

THE UNDERWATER RIVERS
OF THE WORLD

A sailor's geography is not always that of the cartographer, for
whom a cape is a cape, with a latitude and longitude. For the sailor,
a great cape is both a very simple and an extremely complicated
whole of rocks, currents, breaking seas and huge waves, fair winds
and gales, joys and fears, fatigue, dreams, painful hands, empty
stomachs, wonderful moments, and suffering at times.

—BERNARD MOITESSIER
The Long Way

14

The Distant Geography of Water

W E MOTOR THROUGH the clammy mists veiling the coastline. Visibility comes and goes but mostly goes, forcing us to home by sound—the dull thud and whoosh of waves, the piping calls of black guillemots, or sea pigeons, as the local Newfoundlanders call them. We bump in an inflatable Zodiac through shallows and around rocky shoals, taking aboard slaps of water. Now and again an explosion of whale breath detonates nearby, startling us. We can't see the whales, though we can smell their breath and hear the bass whistles played by the sea air across the flutes of their blowholes.

We are laboring under a confluence of conditions—part sea, part air, part land—as tricky as a busy urban intersection at night without benefit of headlights or traffic lights. We are struggling with the wind, the fog, and the slipperiness of the boat, in danger of foundering on submerged objects in all directions. This is the season of the annual migration of icebergs, a spectacle as grand as the exodus of wildebeest through the Serengeti, with the added potential of collisions between us and the migrators.

Newfoundland's drifting nomads calved from their mother glaciers in Greenland two or three years ago, and from that icy world they have meandered south at an average glacierly pace of less than a half-mile an hour, captained by the wind, steered by the currents, following the path of least resistance along a route known as Iceberg Alley. The ones we're seeking to avoid now have survived the ice jams to the north in Baffin Bay, the bottleneck of the Davis Strait, the dead-end fjords of Labrador.

Many lost the race just upcoast of here, inside the labyrinthine claw of Notre Dame Bay. Yet five hundred to a thousand icebergs are still on the move, destined to cross 48 degrees north latitude and dip into the warm waters of the Gulf Stream. A rare few, perhaps two a century, will endure all the way to Bermuda, 1,700 miles south. We are hoping to meet and film a few of these icy nomads, but only from a respectable distance and in clearer air. So we creep, idle the engine, listen, and creep some more.

It's summer, dark, gloomy, cold—*lourd,* as the locals say. The damp and the wind disarm our parkas, and we shiver even in pockets of sunshine. Yet our discomforts are largely forgotten in this numinous theater when white eyelids of fog open to undress the nudes of the landscape: the loins of an Irish-green fjord, a hank of blond kelp bound to the seashore and tongued by waves, the clattered bones of a ghost village sagging to driftwood. In one blink of mist a muscular cliff materializes, tagged with the orange tattoos of crustose lichens hundreds of years old.

And then the curtain opens for good and the sun burns a sideways lantern on the island of Newfoundland and its melancholy set of blue rocks, wind-shorn mosses, black pools, growling waves. Arctic terns (*Sterna paradisaea;* Red List: Least Concern 2008) wheel on the air, their sharp *kip-kip-kip* calls ricocheting from the coves. We take it in, swiveling on the pontoons of the Zodiac, when another eye opens, not in the fog but in the cliff, an arched sea cave the size of Neptune's nose. Although there's much else of interest to investigate, we find our attention, and therefore the Zodiac, reeled into the cavern, into the dark coldness scaled with barnacles and the moist breath of the sea.

We cut the engine and our wake flicks away, climbs the walls, returns, stirring the black water, raising its depths. At first we see nothing, and then we do, just below the surface: milky bells pulsing upside down, as soft as cellophane bags frayed at the tentacles. It's a swarm, or smack, of moon jellyfish (*Aurelia aurita;* Red List: Not Evaluated), hundreds, maybe thousands, of individuals carried into the cave by the tide. They are translucent and then pearlescent, depending on the angle of light coming through the nostril of the cave. Nestled inside the clear tissues of their bells are violet-colored gonads in the shape of shamrocks, some ripe with eggs. Without hurry or direction, these four-leaf clovers, encased in their gelatin bells, tumble in slow motion, jetting as the domes contract, pulsing as they relax, collapsing when two jellies collide, then blossoming open to jet and pulse again, a tireless, mobile heartbeat of the deep blue home.

We can't see it, but the moon jellyfish, for whom travel and feeding are synonymous, are hunting the minute prey of the underwater cosmos—the microscopic planktonic larvae of fish, crustaceans, and mollusks. The pulsing motion of the jellies not only drives them forward—or backward or upward or down—but also drives a current that sucks their prey toward them. This deceptively gentle technique bypasses the sensory alarms of the floating plankters.

It's a graceful scene of death and destruction, this unhurried martial art in a black cave inside Neptune's nose. So we drift, a *rasika* hypnotized by invisible violence. We might flow this way forever, growing stiller and colder—but our dream is interrupted by a splash as quiet as a sniff, followed by an ancient, enormous head, its black skin mottled with white, its outlines smudged, like a charcoal smear, across the imperceptible canvas of black water and pearly medusae. There's no fog in the cave, yet a vapor has lifted, a screen between worlds, and in the blink of a jellyfish's pulse we have looked back through epochs.

It's a sea turtle, and unmistakably a leatherback, with ridges the length of the back, flippers the size of oars, and a head like a draft horse's, wearing a jellyfish mane. We can hear the whistle of the inhale through tiny paired nostrils.

It's not surprising to see a leatherback up here on the edge of the ice, since this is one of the most traveled of all vertebrate species, perpetually on the move along jellyfish highways between the tropics and the high latitudes. Yet it is a wonder to see a reptile at home in this cold realm. And not just any reptile but one bigger than we can imagine, heavier perhaps than the colossal saltwater crocodiles of tropical Australasia. Because of its size, it's likely a male, perhaps nearing his species' known limit of 8.5 feet and 2,000 pounds.

He is big enough to overcome the defining characteristic of reptiles, to become effectively warm-blooded and survive the realm of the icebergs. Here, in water registering between 40 and 46 degrees Fahrenheit, he maintains an internal temperature as high as 78 degrees, a feat managed with the aid of an insulating fat layer and a countercurrent heat-exchange system between the closely aligned arteries and veins in his flippers. The warm arterial blood from his core warms the cool venous blood at his extremities before returning to his heart, keeping the turtle as a whole warm. This warming system keeps his muscles limber and, in combination with the hydrodynamic ridges along his back, enables him to swim faster than other reptiles, in excess of twenty miles an hour, with enough maneuverability to outflank even the swiftest shark predators.

Placid black eyes, half open in nearly vertical lids, observe us. Then he sinks, the black water closing over his oars, the jellyfish moons orbiting the drain of his exit.

· · ·

Moon jellyfish are among the most cosmopolitan of species, eating their way through Langmuir circulations, upwellings, and downwellings, around gyres, along boundary currents, and across global drifts. Technically, they are also plankters—among Earth's largest—wafting principally at the mercy of the currents. Swept by the winds of the sea, they cross an enormous fetch, across a full 110 degrees of latitude between 40 degrees south and 70 degrees north, while tolerating a broad spectrum of water temperatures, from the simmering 88 degrees Fahrenheit of the tropics to the near freezing 21 degrees of Arctic seawater. Except in the Southern Ocean and along the shores of its citadel, Antarctica, moon jellyfish are found everywhere, cinching together the distant geography of water.

Earth's truest nomads (Greek *nomas:* wandering in search of pasture), *Aurelia aurita* are perpetually on the graze while on the move within moving water. It is the simplest existence. Yet like many of the ocean's humblest creatures, they have an exceedingly complex life history, including a peculiar, unmoving, decidedly unjellyfishlike stage wholly different from the graceful medusae swarming inside Neptune's nose.

Sometime in the next week or three, these adults will launch their newly conceived offspring into the currents of the North Atlantic. The males will release their spawn into the water of a smack of jellyfish. The females will suck the spawn into their bells. The fertilized eggs, or zygotes (Greek *zygos:* pair), will spend the next seven to eleven days brooded in the safety of the mothers' specialized tentacles, known as oral arms, and nourished by their own yolks. This maternal investment exceeds that of any other known jellyfish and offers the next generation a significant head start on the difficult path of life. It may also explain, at least in part, the biological success, as measured in abundance, of moon jellyfish.

But before the newly conceived jellies tucked up inside their mothers' bells can become sexually reproducing adults, they must undergo some otherworldly changes. By the time their yolks have been consumed, the zygotes have developed into planulae, tiny flattened larvae that swim away from their mothers on beating cilia to spend the next few days seeking a semipermanent home, preferably in shallow water in the shade. Once attached—or re-cruited—to the bottom, the planulae morph again, this time into filter-feeding polyps resembling anemones. During this sessile phase, the jellyfish are stuck in place as irreversibly as barnacles, fishing for passing plankton with the tackle of their tentacles.

Yet the transformations don't end there. As the polyps mature, they begin to reproduce in intriguing ways, initially budding asexually to form identical new polyps, which stack on top of the original parent polyps like hats on hats. Eventually the topmost polyp, in a metamorphic process known as strobilation (Greek *strobilos:* twisted, turning), asexually produces not another bud but a tiny immature jellyfish medusa known as an ephyra—named for one of the fifty sea nymphs of Greek antiquity. Eventually the ephyra lifts off the polyp stalk and swims away.

Ephyrae look and behave much like miniature adult moon jellyfish, floating and feeding on plankton and dissolved organic matter. Ultimately they mature into adults like the ones in Neptune's nose, and their goal, after as much as a year afloat, is to reproduce sexually.

Success in this final venture depends on the availability of food at the time and place of spawning. In the midst of abundance, moon jellies produce countless small planulae. In the throes of scarcity, they produce only a few larger planulae. Whatever the outcome, the act of sexual reproduction marks the end of the jellyfish's life, and most diminish and die after spawning.

Some version of this remarkably complex life history is played out by all members of the phylum Cnidaria (Greek *knide:* nettle), the marine invertebrates with stinging cells. But most, including anemones and corals, follow a shortened script, with the polyps releasing planula larvae that swim for a while and then settle down to become polyps again, without the free-floating medusa jellyfish stage. One cnidarian, however, the medusa *Turritopsis nutricula,* trumps all others with its startling twist on transformation. After the adult medusae spawn, they do not die but transform back into sexually immature colonial polyps to begin the whole life cycle again. In theory, *Turritopsis nutricula* are capable of immortality.

The ocean is full of secrets tucked into its coves and in the shade on its unseen sea floors, including closely guarded mysteries of longevity. The huge leatherback sea turtle grazing on moon jellies inside Neptune's nose is, if not an immortal, at least an ancient. The truth is we don't know the life span of *Dermochelys coriacea,* though in general the chelonians (Greek *chelone:* tortoise, turtle) are masters of venerability. Harriet, a Galápagos tortoise (*Geochelone nigra;* Red List: Vulnerable 1996), lived to 175 years in a zoo in Australia. Tu'i Malila, a Malagasy radiated tortoise (*Geochelone radiata;* Red List: Critically Endangered 2008) reportedly given to the king of Tonga by Captain James Cook in 1777, survived on that South Pacific island to the age of 188. Adwaita (Sanskrit: one and only), a male Aldabra giant tortoise (*Geochelone gigantea;* Red List: Vulnerable 1996), believed to have been born around 1750 and a pet of the British general Robert Clive in Bengal before taking up residence in the Alipore Zoo in Calcutta, lived to the age of 256.

Many reptiles are long-lived. Chinese alligators (*Alligator sinensis;* Red List: Critically Endangered 1996) are alive and well in captivity after more than 70 years. The sacred crocodiles of India (*Crocodylus palustris;* Red List: Vulnerable 1996) live more than 100 years. A freshwater crocodile (*Crocodylus johnsoni;* Red List: Least Concern 1996) in captivity in Australia is thought to be 130 years old. Henry, a lizardlike tuatara (*Sphenodon punctatus;* Red List: Least Concern 1996) of New Zealand, mated successfully in captivity at 111 years old—middle age for a species expected to live beyond 200.

Other aquatic animals also seem to have remarkable life spans. In 2007 researchers from Bangor University in Wales reported a quahog clam (*Arctica islandica;* Red List: Not Evaluated) of between 405 and 410 years old, born around the time when Shakespeare wrote *Hamlet.* Recent research indicates that some deep-sea black corals live more than 4,000 years, dating back to the construction of the earliest step pyramids of Egypt's Old Kingdom.

Many fish, particularly deep-sea species, including table favorites like orange roughie, halibut, and Chilean sea bass, along with many sharks, reach sexual maturity at 30 or 40 years and presumably live many decades beyond that—implying that those we eat might be more than a century old. Freshwater carp in Chinese fish ponds live beyond 150 years. Pacific rockfish of the genus *Sebastes* are known to exceed 200 years in age.

Marine mammals also exemplify the benefits of the aquatic life. Until recently, most of the great whales were believed to have maximum life spans of 70 years. But the discovery of century-old ivory harpoon points embedded in the bones of Arctic bowhead whales (*Balaena mysticetus;* Red List: Least Concern 2008), along with new calculations based on the biochemistry of amino acids in their eyes, reveal that bowheads can live more than two centuries. One killed by Inupiaq hunters in the 1990s was alive when Thomas Jefferson inhabited the White House.

So the leatherback grazing on moon jellies in Neptune's nose is very likely an ancient, since he possesses several important prognosticators of long life, including large size, taxonomic membership in the class Reptilia, and, most important, residence in the deep blue home. Many decades ago—or even a century or more, perhaps when pirates ruled his home waters—this sea turtle was hatched from an egg buried on a beach in the Caribbean or on the northeastern coast of South America. His gender was determined by the month, and therefore the temperature, in which his egg was laid, with cooler temperatures producing male clutches. Upon hatching, some nine weeks later, he flippered down the beach to the sea and swam hard for six days to reach a highly productive convergence zone offshore, there to consume his own weight in jellyfish every day. Within a few years, gorging himself on meals composed mostly of water along with a few sparse proteins, he managed to grow to roughly 43 inches in shell length—large enough to join other juvenile leatherbacks feeding on inshore jellyfish blooms.

Part of his ongoing success surviving on such meager nutrition lies in the specialized esophagus running the length of his body and doubling back toward his mouth—an esophagus he stuffs sausage-full of jellyfish hour after hour, day after day, decade after decade. In pursuit of jellyfish he swims nonstop, following migratory corridors along deep undersea topography, navigating along cryptic currents and weather fronts, using the maps of shorelines and the pull of magnetic fields. In his lifetime he has completed grand transoceanic circuits and tangential jaunts along a fairly predictable schedule: off Florida in September; off Virginia in April; in the Gulf of Maine in June; perhaps some years to northern Europe; perhaps a tour to the Indian Ocean.[1] Of all the itinerant sea turtle species, leatherbacks travel the farthest.

15

The Ecumenical Sea

IN THE NINTH AND TENTH *Mandalas,* or cycles, of the Rig-Veda, *Rasā* is the name of a mythical river flowing around the world, uniting the land, the sea, and the air. *Flow round us with thy protecting stream, as Rasā flows around the world.* This moving swirl bestows life upon the lifeless, transforming our world from a body in space to a heavenly body endowed with the energy of birth and renewal.

The ancient Greeks reworked *Rasā* into their own interpretation of a uniting, circulating life force, which they personified as Oceanus, ruler of the ocean. Like *Rasā,* he unified sea and air to create the *oikoumene* (from which we get *ecumenical*), or inhabitable world. Like *Rasā,* Oceanus was older even than the gods, a member of the Titans, or giant elder gods, son of Uranus the Sky and Gaia the Earth.

A millennium after Oceanus, Norse sailors braving the iceberg alleys of the North Atlantic en route to the world they would call Vinland bore with them their own belief in a primeval life-bestowing deity of the sea—Ægir, likewise an ancient, a giant of the race known as the *jötnar*. Some Norse sagas portrayed Ægir as the oldest of beings, older than the gods or the *jötnar*, a mysterious deep-rooted force whose beginnings could not be known.

Ægir's hall was said to be lit by floors strewn with gold, and for this reason the Norse referred to gold as the *fire of the sea*.[1] His sanctuary was so warm and inviting that the younger gods arrived in their splendor each winter to avail themselves of his magically refilling tankards of beer. Ægir's consort was the sea goddess Rán, keeper of the drowned, who on bad days ensnared sailors in her nets. The pair's watery primordial couplings produced nine daughters, the billow maidens, veiled in white and named for the nine forms of waves, from gentle to wrathful: Himinglæfa (Transparent-on-top), Dúfa (Wave), Blódudhadda (Bloody-hair), Hefring (Lifting), Unn (Wave), Hrönn (Wave), Bylgja (Billow), Kára (Powerful), Kólga (Cool-wave).

Off Newfoundland this afternoon, the billow maidens are asleep, dreaming under their blanket of fog. We nose the Zodiac through still air and calm water, the outboard gurgling, its exhaust cloud hovering behind us like an unwelcome guest on the stern. We expect the day to be a changeling, as all days are in this boreal world, with fog and wind and rain and sleet coming and going as Ægir tosses and turns and throws off his blankets, while Rán, good *húsfreyja*, washes them, wrings them out, and snaps them dry in the wind.

We are motoring through their insomnolent house when out of the fog comes a patch of blue sky and, sailing though its center, an iceberg so blinding it makes us squint, so close it makes us lean away. Its skin is wind-polished to the luster of diamond, water-sculpted to sapphire, its form purely feminine, purely voluptuous, curves hewn hard in the foundry of the far north. Although the air temperature is cold enough to make us shiver, it is not cold enough for this relic of snowfalls past to survive, and a silver cascade of meltwater pours from its summit down a chute of polished translucent turquoise. Crystal by crystal, its ice diminishes, collecting in a pool on the iceberg's saddle, where Arctic terns dip and splash, unfurling freshwater rainbows.

Deep blue veins and track marks of gravel scar the ice, records of good years and bad years in the glacial world. Somewhere in these layers are the remains of blizzards, circa 1000 CE, when Leif Erikson, son of Erik the Red, left Greenland and tried and failed to found a colony in Vinland. Somewhere in this white book of ice is the meteorological memory of 1010 or so, when Thorfinn Karlsefni, sailing from Iceland with three ships and 160 settlers, succeeded for three brief years in inhabiting the New World.

The Vikings settled at a bay in the north of the island of Newfoundland named by nomadic nineteenth-century French fishermen L'Anse aux Méduses, or Jellyfish Cove. The French name was later corrupted by Irish and English settlers to L'Anse aux Meadows in recognition of the pretty shoreline ringed with grassy hummocks. In 1960 archaeologists excavating the hummocks found the remains of eight buildings dating to about the year 1000, since identified as the first known Norse settlement outside of Greenland, perhaps the legendary Vinland itself.

Like jellyfish on their mysterious peregrinations and leatherback sea turtles on their epic swims, Viking sailors of old, with the aid of winds, currents, and stellar navigation, found their way into these jellyfish-rich waters, where leatherbacks, the ancestors of the enormous turtle in Neptune's nose, gathered to gorge amid the ice.

We pull alongside the sunlit iceberg with its waterfall trickling from the ice bowl on its saddle into the sea. The Vikings made their voyages here at a time when the North Atlantic was freer of ice than it is today and the climate warm enough to support their livestock in Greenland and Newfoundland. Somewhere in this floating remnant of the glaciers lie passages from those Norse lives, remnants of their wood smoke or dung smoke, wool fibers from their sheep, traces of human hair.

But the naked eye is not discerning enough for such far-off tales. So we scan the ocean for another story. The fresh water from the melting iceberg swirls on the surface in an oily-looking lens. Air bubbles freed from the melting of submerged ice rise and bob against this freshwater cap, dimpling the surface, as if rain were falling upward from underwater. The illusion is reinforced by the spritzing we receive from the bubbles bursting and exhaling their 15,000-year-old air.

As quickly as we can in the cold, we change out of our parkas and into our wetsuits. It's 1984, and my partner in our newborn documentary business, Hardy Jones, and I are trying to become film professionals. This is a vocation without textbooks or compasses, and our transition involves traveling the world disencumbered by the customary hallmarks of professionals, like budgets and contracts. Instead we cling to hope, that seven-tenths submerged vessel, which will either sink us or float us to the far-flung regions of our dreams.

We are also disencumbered by gear. We have neither exposure suits for topside work nor drysuits for underwater work. Therefore we're cold before we even enter the water. Slipping overboard in our ill-fitting neoprene, cameras in hand, we are smacked with the full-strength experience of the North Atlantic—a wave of icy water not cold enough to keep an iceberg frozen yet frigid enough to ice-pick through us to the marrow of our bones.

It doesn't matter, though, because the underwater world is like a jolt of electricity, the sea fizzing with bubbles aglitter in a floodlight of ice. The sides of the iceberg are furred with newly born bubbles cavorting upward in trembling clouds—half-formed creatures that break apart and disintegrate against the ceiling of the sea. Mists of bubbles envelop us, tickling the bare skin of our faces and whitening the undersides of our wetsuits. It's cold. But intoxicating. Like diving in Ægir's aquamarine beer.

Nor are we alone. An enormous shadow looms below. We strain to make it out within the veil of bubbles—a shadow and then not a shadow. It's one creature and then not one but thousands of creatures—a shoal of silvery fish, the sardines of the north, known as capelin (French Canadian *capelan:* chaplain; *Mallotus villosus;* Red List: Not Evaluated). They huddle in the center of the bubble curtain, hiding.

From what, we're not sure. And then we see, shooting up from below, two hooded seals (*Cystophora cristata;* Red List: Vulnerable 2008), speeding through the carbonation on the propellers of their hind flippers. Their splotchy silver-and-black pelage paints a natural camouflage in the swirl of bubbles. The capelin know all about hooded seals, however, and recombine into one organism, flashing their shiny bellies before disappearing into the smokescreen of their darker backs, fighting to elude their predators with weapons of optical illusion.

The seals know all about optical illusions, however, and torpedo into the smokescreen. The shoal disintegrates into a thousand shards of fish, and the fight spirals downward in a silvery tornado. We free-dive after it, descending the ice-haunches through the darkening beer, where the groaning voice of the berg—a low woodwind interspersed with rifle-fire pops—sounds a warning, perhaps from the formidable Rán herself. The drama fades into the shadows beyond our reach, and we return breathless to the surface. Our last sight is of the sparkling fish merging with the splotchy seals until we can no longer tell where one species begins and the other ends.

We lurch, frozen, into the Zodiac, gasping on our backs like hooked fish . . . though our disappointment is soon assuaged by the appearance of a minke whale (*Balaenoptera acutorostrata;*

Red List: Least Concern 2008). This sleek wanderer of the ice realm is the smallest of the great whales, at only twenty-five feet in length and five tons in weight. It is a graceful presence, coursing through the bubbly sea on lithe flukes, curious about us but difficult to approach and quick to disappear, as coy as a sea nymph.

We try for some topside shots. But the magic of this species includes its contortionist ability to transform itself on film into something worse than invisible, its dark gray dorsal fin skipping through gray wave tops like an unwanted piece of dust on the negative. Minkes never throw their flukes into the air when sounding but only tease, rolling with an arched tail stock and no flukes. But we know that they do occasionally breach—that is, throw themselves completely out of the water—and that they sometimes do this repeatedly. So we follow from a hopeful distance, even though breaching is hard to catch on film, particularly at a time when documentary budgets are calculated, like lumber, by the board foot of exposed film.

The minke is also hunting capelin, and if we're lucky we might see it course up through a ball of fish, jaws wide, gulping. That would be worth at least an entire 400-foot magazine of 16-millimeter film. Ægir's daughters are quiet enough at the moment to inflate this hope further, allowing us to peer through the surface and see the whale's pretty countershading—dark gray above, white underneath—a capelinlike ruse that matches the whale with the sky when seen from below, and with the darker abyss when seen from above, hiding it from both predators and prey.

We are also hoping to see the seals again, so we drift farther from the iceberg, widening our circle of possibility. While we wait, Arctic terns fly to and from the meltwater pool. The bubbles rise and pop. The whale stubbornly, unbreachingly remains beyond useful camera range. While we wait, Rán awakens, stirring the wind and effectively dropping the temperature by 20 degrees, slapping the waves into *lops,* as the Newfoundlanders call them, small breaking seas that splash into the Zodiac. Still we wait, wet, shivering, blue, as Guy, our French Canadian companion, regales us, teeth chattering, with stories of diving on icebergs, of entering their secret underwater caves and finding seals inside sleeping on dry ledges.

We are bitterly cold in more ways than one when we abandon the day and make landfall to haul the endless cases of camera gear up the long hill toward the car. We are frozen and wet and feeling sorry for ourselves about the missed seals, the missed underwater sleeping caves, the capelin mob, which doubtless the minke whale is feasting on right now. Our misery, internal and silent, does not prevent us from hearing the monstrous, rumbling detonation in the bay behind us. Surely some war has begun, we imagine, in that instant when we turn, wet neoprene booties squelching, to see the iceberg turning upside down, its huge underwater keel slicing like a blue fluke into the air and listing in the wind before disintegrating into a debris field of slush and brash ice skidding across hundreds of yards of ocean surface.

In less than one minute, this far traveler from the far north, this *jötunn* from ice ages past, has disappeared from being, relinquishing its molecules back to the ocean—one less ancient to animate our giantless age.

16

Deepwater Formation

TWO THOUSAND YEARS after the Titan Oceanus ruled the waves, modern oceanography confirms what the Greeks knew: that the globe is indeed encircled by an ocean river far more powerful than any Nile or Amazon. This Möbiuslike ribbon rises and falls, stirring the waters of the world from their sunlit surfaces to their abyssal depths, a *Rasā* combining the power of atmosphere and ocean, assembling waves in air and water that build and unbuild shorelines, carrying even to the landlocked hearts of the continents, keeping Earth wet and alive.

In modern parlance *Rasā* is known as the ocean conveyer belt or, technically, the thermohaline (Greek *therme:* heat; *halos:* sea salt) circulation. Whatever the name, it is the deep vascular system of our planet. Like the leatherback sea turtle's countercurrent heat-exchange mechanism, it brings warmth to the cool extremities of Earth, uniting our arbitrarily divided oceans—Atlantic, Pacific, Indian, Arctic, Southern—into one *oikoumene*, one World Ocean, empowering the atmosphere, watering the land, and enabling life as we know it.

The power of the thermohaline circulation is deep and ele-
mental, godlike, if we were still inclined to think that way. Its
might derives from profoundly slow-moving currents, distinct
from the powerful surface currents that affect only the top 10 percent of the ocean down to about 660 feet. The underwater rivers
of the world, segregated from each other by ranges of chemistry
rather than rock, by the contour lines between different temperatures and degrees of salinity, the ingredients of density, run far
below our sight, yet not beyond our ken.

Just as gravity drains the rivers of the land, so gravity drains
the rivers of the World Ocean. The salticr, colder, heavier rivers sink beneath the fresher, warmer, lighter ones. The three-dimensional realm of the ocean is layered with watersheds running
over and atop one another in multiple directions. An exploded
view of the global thermohaline circulation looks something like
an intricately entwined highway interchange system, with layers
crossing and bypassing at many levels, in all directions, and at
different speeds.

The Gulf Stream, for instance, the swiftest surface current
on our planet, is driven mostly by wind and carries warm water
from the tropics to the Arctic. Yet directly below the Gulf Stream,
the Deep Western Boundary Current, driven by gravity, flows in
exactly the opposite direction, south toward the Equator. These
two components of the thermohaline circulation are composed
of some of the same waters, separated by time.

The changeover between these two currents occurs near the
northern reach of the Gulf Stream's range, around Greenland,
Iceland, and Norway. In these high latitudes, the warm surface
waters shed their heat to the cold, windy Arctic air—incidentally
warming Europe. As the water cools, it also evaporates, resulting
in fewer water molecules in relation to salt. This cooler, saltier
Gulf Stream, now too dense to remain afloat, dives down, a process known as the meridional overturning circulation—a trans-

dizzying
fallen
up.
(st
x)

formation as profound as that of a moon jellyfish polyp becoming a medusa.

The consequence of this vertical overturning is the deep mixing of the waters of the World Ocean on a time scale more *jötunn* than human. Drops of rainwater falling off Florida flow north via the Gulf Stream toward the Arctic, growing colder, sinking, and turning to flow south—some via the Deep Western Boundary Current, but more through a newly discovered and as yet unnamed central pathway around Newfoundland. By whichever route, these cold waters run thousands of feet below the surface toward the subtropics and farther south, eventually converging with the densest water mass on Earth, the Antarctic Bottom Water. This supersalty water plummets in a vertical current and seeps along the sea floor north into the deepest basins of the Pacific and Atlantic oceans, reaching the equatorial Pacific in a thousand years and the Aleutian Islands in another six hundred years.

Deepwater formation occurs at only a handful of places on the globe, including the Greenland-Iceland-Norwegian Sea, the Labrador Sea, the Mediterranean, the Weddell Sea, and the Ross Sea. Each of these formations is a potent, almost omnipotent, component of our global weather system. The Little Ice Age that struck Europe and North America circa 1300 may have been caused by the faltering or failure of the North Atlantic meridional overturning circulation, brought on by the Medieval Warm Period and enough melting Arctic ice, perhaps, to dilute the salinity of the North Atlantic and redirect the flow of the underwater rivers.

Whatever its true cause or causes, the resulting Little Ice Age profoundly affected human endeavor in Europe and the New World. Rivers and harbors froze in winter. Glaciers advanced in the Alps, entombing villages. Inuit people were seen paddling along the ice off Britain's coast. Cold and damp conditions fueled the spread of the Black Death, which killed as many as three of five Europeans. In the northern countries, growing seasons declined below sustainability, forcing farmers to abandon once productive lands—particularly in Greenland, where the long Norse tenure was terminated by cold.

We can deduce the struggles of these Vikings in their deteriorating western outposts, with shorter growing seasons and increasingly ice-clogged trade routes back to Iceland and Norway. We can imagine how the last Norse in North America endured great cold and hunger, how they prayed in their fine Christian church at Hvalsey, perhaps yearning for the days when the bounties of Ægir's golden hall warmed their bones.

17

The Tempest from the Eagle's Wings

THE GREEK BOREAS, winged god of the north wind—from whom we get the term *boreal*—was said to inhabit the world of Hyperborea, the place beyond the north wind, a realm where people lived uncommonly long lives in a state of unusual bliss. His Norse counterpart, Hræsvelg, was portrayed as an enormous sea eagle who drove the north wind by fanning his colossal wings. The Icelandic poet Sæmund the Wise wrote of him:

> *Where the heavens' remotest bound,*
> *With darkness is encompassed round,*
> *There Hræsvelg sits and wrings,*
> *The tempest from his eagle's wings.*[1]

At this moment, on Newfoundland's southern shore, Hræsvelg is flapping hard, laying flat the white heads of cotton grass and blowing the grazing sheep into the steep flanks of the hills, tilting them onto the edges of their cloven hooves so that they appear to be listing under woolly sails. Hardy and I are shivering under this wind but also sweating as we labor on foot across spongy meadows, carrying heavy cameras and tripods, dodging tufts of wool and a blizzard of feathers driving past us on pungent scents of barnyard and seascape and the incongruous sardine-can breath of humpback whales.

A trio of whales is feeding close inshore, yet so far below us that we are afforded an aerial view of their gray-blue bodies, fringed with white on flukes and pectoral fins. We are here to film humpback whales (*Megaptera novaeangliae;* Red List: Least Concern 2008), though we have not found any until today. And now here they are, swimming through rough seas far from a boat ramp on a day we don't have access to a boat.

The whales, surging across the surface, are churning it to froth. Hræsvelg is shooting arrows into their wakes from the drawn bow of his wind, in the form of seabirds known as northern gannets (*Morus bassanus;* Red List: Least Concern 2008). These masters of the boreal seas plunge-dive from on high, spearing powerfully through the surface while somehow managing to avoid the many members of a floating mob of black-legged kittiwakes (*Rissa tridactyla;* Red List: Least Concern 2008). Huge flocks of these tickleasses, as the Newfoundlanders call them, are paddling expectantly in places where a whale is destined to rise from a dive. From our vantage on the bluffs, we can see that they have expected correctly, that a rising whale's whitened snout is lighting the way from the deep directly toward the undersides of the tickleasses and the lancelike bills of the diving gannets. The whale is driving a thin school of fish toward the surface, an hors d'oeuvre, not a meal.

The humpbacks are here to hunt capelin and the birds are here to hunt capelin and we are here to film capelin and whales, but the capelin have not arrived in their expected numbers. Part of our labor in becoming filmmakers is arriving at our destination at the wrong time. Not because we have not done our research but because nature's timetable is tricky and because, on a purely practical level, we sometimes wrangle free airline tickets with expiration dates incompatible with nature's timetable. Nevertheless, whatever we miss because we caught the wrong plane is invariably replaced by whatever we would have missed had we caught the right plane. As Ajahn Chah, the Thai Buddhist master, wrote: "We have to understand the way things really are, the way things contact the mind and how the mind reacts, and then we can be at peace."

What has contacted our minds just now is the cold beauty of a landscape that we would not have known had the capelin been spawning, had the whales been hunting them, had the whales been inadvertently getting entangled in the fishermen's capelin nets, leading our host, Peter Beamish, to attempt heroically to cut them free. But because none of this is happening, we have been forced to improvise by driving halfway across the island in Peter's battered station wagon—a trip enlivened by a near collision with a semitrailer on a two-lane road in dense fog. Peter is an enthusiast, in the original sense of that word: "one who is (really or seemingly) possessed by a god; one who is under the influence of prophetic frenzy," according to the *Oxford English Dictionary.*

We have survived the overland expedition to witness now the magnificent 300-foot-high bluffs along the Avalon Peninsula, that fjord-scarred pendant dangling from the southeasternmost shore of the island of Newfoundland by a three-mile-thin isthmus of land. The Avalon Peninsula is guarded by cliffs and the sea—and not just any sea, but the plush bed of the continental shelf known as the Grand Banks, the incomparable fishing grounds that produced enough high-octane fish to power human civilizations on both sides of the Atlantic for half a millennium.

The Grand Banks also fuels some of the largest and most enduring seabird colonies on the eastern seaboard, including Cape Saint Mary's, the Isla Rasa of Newfoundland tilted on edge. Seabirds crowd every square foot of the cape's perpendicular cliffs and sea stacks, clinging to the slimmest toeholds of rock and moss with the tenacity of bats: northern gannets, black-legged kittiwakes, common murres, thick-billed murres, razorbills, black guillemots, double-crested cormorants, great cormorants, northern fulmars. *Cape Saint Mary's pays for all,* say the fishermen of Newfoundland in tribute to the region's riches of fish and birds.

The choicest real estate on the cape is Bird Rock, a 350-foot-high sandstone sea stack rising precipitously from Saint Mary's Bay and separated from the coastal bluffs by a 60-foot chasm of air. Seen from afar through veils of fog and across green meadows popping with wild purple irises, Bird Rock resembles nothing so much as an outrageous lemon meringue cupcake in the process of being blown apart by the wind, its frothy whites and yellows spitting into the sky and falling back to the rock or plummeting to sea.

Of course it's not a confection of whipped eggs and sugar but a colony of northern gannets, who own the domed heights of Bird Rock simply because they are the largest seabirds in the North Atlantic, armed with six-foot wingspans and formidable spearlike bills colored pale blue and rimmed with dark blue-gray lines. No other seabird can birdhandle the territory from them. From a cinematic point of view, gannets are the natural owners of the high ground by dint of their beauty alone: their snowy plumage crowned with golden yellows on neck and head, their pale blue eyes ringed in cobalt-blue kohl. They are stars behaviorally too, performing noisy, intricate, camera-friendly courtship displays: pairs standing breast to breast, wings out, heads shaking, fencing bills together with loud knocks, while croaking and chuckling.

. . .

In 1984 this confluence of land, water, and birds is enjoying only its second year of protection as the Cape Saint Mary's Ecological Reserve—a full twenty years after Mexico protected Isla Rasa by federal decree. The lag tells something of the unhappy relationship between Newfoundland's fishermen and Newfoundland's seabirds—and something of the relationship in many places between fishermen and seabirds, who often pursue the same prey.

For centuries Newfoundland's fishers hunted nesting gannets, whose flesh they used as bait for codfish—an economy of action that amounted effectively to killing a fish-eating bird and a fish with one stone. In his *Life Histories of North American Petrels and Pelicans and Their Allies,* Arthur Cleveland Bent described how a half-dozen club-wielding Canadian fishermen of the nineteenth century could lay waste to five hundred birds on the nest in the space of an hour, slugging the birds senseless and tossing their bodies into boats far below. Farley Mowat wrote that Newfoundland's twentieth-century fishers improvised by using capelin to bait partially sunken waterlogged planks, then setting them adrift near Bird Rock to booby-trap the diving gannets into breaking their necks.[2] Even today, given the opportunity and sufficient privacy, many fishers will shoot or hook gannets or catch them by hand if they land on or near fishing boats and toss them back to sea with the head or a leg tied to a wing or with the bill bound with rubber bands.

In the collection of Ajahn Chah's Buddhist teachings titled (in lowercase) *everything arises, everything falls away,* the great master tells a fish story. There was once a man who loved fishing and could not stop despite the Buddha's teachings. The fisherman came in search of Ajahn Chah's instruction, and after a while he reached the point where he would drop a hook into the water and proclaim aloud that any fish which had reached the end of his karma could come and eat his hook, and those which hadn't should stay away. Ajahn Chah continued his instruction, and after some time the fisherman could look a fish in the face and see the hook caught in its mouth and begin to feel something of its pain. Yet he told himself that he had warned the fish not to come to his hook, and what could he do if the fish persisted?

The eastern seaboard of North America once hosted scores of gannetries, some bigger and far more populous than Newfoundland's Bird Rock. By the middle of the nineteenth century, only nine survived, and eventually this number was whittled to six. John James Audubon visited the largest of these, Québec's Bird Rock, in 1830 and wrote: "The air above for a hundred yards, and for some distance around the whole rock, was filled with gannets on the wing, which, from our position, made it appear as if a heavy snow was directly above us."

Thirty years later the naturalist Henry Bryant estimated that 150,000 birds populated the Québec colony. Yet only twelve years after Bryant's visit, in the aftermath of the construction of a lighthouse, the ornithologist C. J. Maynard counted less than 5,000 gannets.

Everything arises, everything falls away. And sometimes everything arises again. Because even as the number of gannets was in steep decline in the late nineteenth and early twentieth centuries, the first portable prism binoculars appeared on the market. Carl Zeiss's new tools enabled people increasingly unfamiliar with nature to call into closer view the natural world, be it far-flung sea stacks or backyard elm trees. Suddenly, through binoculars, an inaccessible snowfield crystallized into captivating views of individual birds with six-foot wingspans and formidable blue bills. Miraculously, a hum of blurry motion focalized into pairs of birds performing intimate displays, wings out, heads shaking, knocking bills together, the birds' blue eyes as close and familiar as the eyes of a lover.

When Ajahn Chah's fisherman arrived at the point where he could truly renounce fishing, he was ordained as a Buddhist monk with Ajahn Tongrat. What must I do? he asked. Sit by this old tree stump, said Ajahn Tongrat, the silent master, and make yourself like it. And so, through the prisms of stillness and contemplation, the ex-fisherman developed intimacy with the real world and took as his lover the way things really are and began to find the way home.

18

One Meritorious Act

THE RISE OF THE SILENT MASTERS, binoculars, coincided with the decline in the nineteenth and early twentieth centuries of birds falling victim to the peculiarly extravagant Victorian hat fashions designed in theatrical admiration of nature, albeit at nature's expense.

When the ornithologist Frank Chapman of the American Museum of Natural History strolled Manhattan sidewalks on two days in 1886, he spied forty species of North American birds adorning three quarters of the 700 hat-wearing women who passed by. The adornments included feathers, nuptial plumes, whole wings, whole tails, even whole birds, stuffed and mounted on oversized Beach Blanket Babylonian headgear. Chapman's feathered hat census recorded grebes, jays, herons, bluebirds, rails, robins, yellowlegs, shrikes, sanderlings, thrashers, gulls, terns, waxwings, warblers, grouse, prairie chickens, bobwhites, sparrows, quails, doves, owls, flickers, meadowlarks, woodpeckers, grackles, orioles, tanagers, flycatchers, grosbeaks, and swallows.

Eventually sixty-four species in fifteen genera of native American birds were victimized by the feather trade. A hundred years after Chapman's census, Paul Ehrlich, David Dobkin, and Darryl Wheye cited the damage from a single London sales source in 1902: 1,608 packages of herons' plumes, weighing in at 1.5 tons and representing some 193,000 herons killed at their nests, with two or three times that number of nestlings or eggs incidentally destroyed.

Fourteen years after his Manhattan census of behatted ladies, Frank Chapman published a letter to the editor in the *New York Times*. The public, beginning to feel uneasy about the downstream effects of feathery fashions, was attempting to allay its disquiet with the sedative of misinformation. Chapman wrote to correct a commonly held belief of the time—disseminated in part by the 1-in-1,000 Americans employed in the feather trade—that most millinery plumes came from domestic chickens or game birds:

> I would suggest these same feather workers visit the American Museum some Sunday afternoon in the quest of a little ornithological information. They will discover that grebes, gulls, terns, egrets, buzzards, eagles, hawks are neither domestic fowl nor game birds, and doubtless on returning to their work the following morning they will recognize among the so-called "chicken feathers" about them the plumage of most of the birds observed in the museum the preceding afternoon. If additional evidence be required we can supply it, we regret to say, in unlimited quantities. We not only have the original orders sent to the hunters by the feather dealers, but we have the confessions by the hunters of tens of thousands of birds killed in a single season, and when we visit localities where certain birds were once abundant, their comparative scarcity, or entire absence, is testified to by our own eyes.

Chapman, of course, was fighting a losing battle against a situation impossible to solve, with too much money, too many jobs, too much international trade tied up in a business that could not be dismantled.

Yet it was. Four forces converged upon the problem. The first, public education, was spearheaded by the likes of the newly fledged Audubon Society, which presented lectures titled "Woman as a Bird Enemy." The second, the supply of feathers, was affected by the alarming decline of the feather suppliers, the birds themselves. The third, regulations and restrictions, were imposed by the much-maligned hands of governments, which came to the defense of the defenseless. Federal laws were enacted criminalizing interstate commerce in birds, and eventually such laws flowed across international boundaries. The Migratory Bird Treaty of 1918 was driven by the concerns of birdwatchers on both sides of the Atlantic, who endeavored to convince Britain, on behalf of its dominion, Canada, to sign this revolutionary agreement protecting birds and eggs from commercial exploitation. The treaty was enacted against all odds in the midst of World War I—the fourth force converging upon the feather-trade problem by stripping excesses from a formerly profligate society.

Everything arises, everything falls away.

When the mind is settled and still, wrote Ajahn Chah, *wisdom will be able to see things.* When enough minds are settled and still, wisdom will be able to act.

The Migratory Bird Treaty of 1918 and its enabling national legislations in Canada and the United States gave the surviving gannetries of the eastern seaboard the buffer they needed to fight their way back from the edge—no easy task for K-strategists, who mate late and produce small clutches.

But the birds cooperated with the full intensity of their life force. They met on their ancient gannetries each season, seeking mates, either new ones or life mates from seasons past—the partners who would put the *ornaments of their bodies* upon each other. Pouring their energies into the one egg produced per season, each gannet pair spent the six months between March and September at familiar nest sites, building or repairing with seaweed, grass, molted feathers, straw, guano. Beachcombing males collected an eclectic inventory of northern beaches, including—according to two ornithological tallies—netting, rope, fishing line, plastic wrapping, lobster-pot tags, shotgun-shell casings, a plastic frog, false teeth, a catheter, a gold watch, a fountain pen, golf balls.

Gannets fished to live and lived to reproduce. They foraged up to 110 miles away at sea during the breeding season and spread even farther into the North Atlantic in the nonbreeding season. By 1939 the British naturalist James Fisher estimated that the six remaining gannet colonies of the western North Atlantic had recovered to 28,000 breeding birds. When he visited Cape Saint Mary's in 1953, in the company of Roger Tory Peterson, the American naturalist and inventor of modern field guides (the next generation of silent masters),[1] both men proclaimed that gannets were on the increase.

In 1984, as we set our cameras on Newfoundland's wind-torn bluffs, the colonies are beginning to recover from a second, albeit invisible, clubbing—from the same pesticide assailants that felled peregrine falcons in North America. It's twenty years after Rachel Carson's impassioned plea for rationality: "This sudden silencing of the song of birds, this obliteration of the color and beauty and interest they lend to our world have come about swiftly, insidiously, and unnoticed by those whose communities are as yet unaffected."[2]

Like peregrines, gannets are large birds who eat large prey and who have in turn accumulated high levels of whatever chemicals are afflicting their world—a process known as biomagnification. Once absorbed, these persistent organic pollutants are difficult if not impossible to offload. Many high-trophic-level predators (Greek *trophe:* food) — the animals who feed on the upper levels of the food web, including ourselves — carry far higher concentrations of these chemicals than exist in the surrounding landscapes and waterscapes.

Nowadays persistent organic pollutants have infiltrated the deep blue home to such an extent that the bodies of some top predators meet our definitions of toxic waste — including some dolphins in the Mediterranean, whales off Washington State, harbor seals in the North Atlantic, and beluga whales in the Gulf of St. Lawrence. One of the more insidious effects of this brand of pollution is that it skews the sex ratios, reducing the number of males. Female marine mammals are able to offload contaminants in their bodies through their breast milk, lessening their own toxic load, though poisoning the children. Among some killer whales in the northeastern Pacific, reproductive success is declining as calves die, particularly firstborns, while adult males are perishing years younger than females, presumably because they have no means to offload their own toxins.

Pollutants efficiently ride the rivers of the World Ocean to contaminate waters far from their source points. In parts of the Arctic, polar bears (*Ursus maritimus;* Red List: Vulnerable 2008) carrying high contaminant levels are breeding poorly and faring poorly because of compromised immune systems. The people who rely on the Arctic food web are suffering too, with persistent organic pollutants accumulating in the tissues of Greenland Inuits enough to suppress the immune system and levels of vitamin A. The Arctic Monitoring and Assessment Program cites evidence of neurological damage in human fetuses, as well as decreased resistance to infections in the first year after birth.

In 1984, twenty-two years past Carson's song for a dying world, we focus our camera lenses on Bird Rock. The gannets are recovering from a ban on DDT. Their eggshells are thickening and the hatching rate is on the rise. Although other pollutants are work-

ing their way through the pipeline of the River Rasā and are destined to reach the birds on a future circuit of currents and winds,
at this moment Cape Saint Mary's breeding gannets have arisen
to more than 10,000 birds, with a total of 79,000 gannets in the
surviving rookeries of the western Atlantic. This trend informs
us that adversity, too, can fall away.

Filming from the bluffs, we are buffeted by the collective voice of
the birds, as rough as the sea below and likewise rising, cresting,
breaking into squawks, chuckles, squalls, trills, shouts. Dr. Charles
W. Townsend, visiting the gannets in 1919, described their calls as
"the sound of a thousand rattling looms in a great factory."[3]

Thousands of long-winged, long-tailed birds are woven onto
the wind, white crosses billowing on veils of fog. Heads down
or cocked to the side, the gannets call to their mates below, sift-
ing through the mayhem for familiar return calls, their personal
homing beacons. When heard, the receiving bird retracts its
wings, extends its legs, and drops—as light as a falling handker-
chief that neatly folds itself upon landing.

The great looms of the gannets are joined by other calls: the
breaking-glass *kit-tih-wake!* of the kittiwakes; the purring and
moaning of turrs (Newfoundlandese for murres); the grumpy
growling of razorbills; the piping and hissing of guillemots; the
strangled croaks of cormorants; the adenoidal cackling of fulmars. Add to these the high-pitched peeps of chicks, including
those still in the egg yet already talking to their parents, and you
have a full-fledged United Nations of multispecies languages—as
if all the tongues you hear on the New York City subway were
not just those of humans but also those of our relatives—goril-
las, chimpanzees, bonobos, orangutans, gibbons, macaques, ba-
boons—complete with their children in utero and ex utero . . . a
loud, loquacious, repetitive, insistent, hurried, harried, and vital
discourse.

We capture as many dramas as we can crowd onto our limited

allotment of film: the courtships, the copulations, the pipping of eggs, the chicks diving into their parents' gullets, the parents diving into the thrashing waters, flight in all its varieties of aerodynamic design, stretched on the stop-motion of the wind. With long lenses we investigate the nests of diving birds on the cliffs below in search of interesting finds from the deep. To the north, off the coast of Labrador, fishermen report nests full of pocketknives, smoking pipes, hairpins, and ladies' combs, salvaged by cormorants diving onto the wrecks of old trading vessels.

Each day we should do at least one meritorious act, wrote Ajahn Chah. *At the very least you can show kindness to an animal.* Like traveling photographers of the nineteenth century, we set up our heavy cameras and point them at the winged locals dressed in their best plumed finery in the midst of their busy lives, hoping to capture their portraits for posterity, should posterity survive. In 1984 we are hopeful our pictures might contribute to everything wild arising a while longer.

19

Jump Cut

TWO GREAT SURFACE CURRENTS converge in New-foundland's waters, the warm, north-flowing Gulf Stream and the cold, south-flowing, iceberg-laden Labrador Current. The confluence breathes a humid fertilizer into the atmosphere, growing fogs as thick as milk, which flow profusely over this boreal land and boreal sea. Fog is Newfoundland's norm, drying only briefly when wind and sun flex their muscles. The Grand Banks is said to be the foggiest place on Earth, with the shores of Newfoundland running a close second. Cape Saint Mary's clocks more than two hundred foggy days a year.

We are sailing the verge of the banks, at the outer edge of Trinity Bay, aboard Peter's little sailboat, *Ceres*, trying to find whales in the fog, rarely a profitable occupation. We have seen in the distance, thrown up on the white screen of a far iceberg, the mighty black flukes of a sperm whale, evidence of two tantalizing realities: first, that in these high latitudes the flukes belong to a large, solitary bull whale, of which not many remain in the deep blue home; and second, that he has sounded and will probably be hunting below for an hour.

So we resort to shooting what in the film business is called cutaways, the filler shots inserted between pieces of film that otherwise won't connect. Cutaways are useful to bridge jump cuts, those unpleasant reminders facing us weeks later at the editing table of our failure to shoot something well in the field—such as the breaching whale who flung himself out of the water and pirouetted a pretty 180-degree barrel roll before falling to the surface on his back. Of course we missed the beginning of the breach, since we were looking the other way when it started. What we did catch was an image of a whale miraculously and unaccountably in midair with no palpable explanation of how it got there.

We could mask the jump cut in the editing room with sound effects, a splash, for instance, over a picture of empty water, leading to the shot of the airborne whale. But far better to insert a cutaway between the two mismatched images. We might use a shot of a gannet afloat on the air, though it can't be just any gannet afloat on any air. It must be a gannet in roughly the same foggy air, and preferably in close-up, suspended on the sails of its wings, head down, looking at something, then turning its head sharply just as the whale lunges from sea to air. The gannet's shifting attention creates a seamless, albeit fictional, transition, convincing enough that viewers, ourselves included, will believe that the whole breach unfolded before their eyes.

Thus we are hoping for an airborne gannet to aid us in our desire for a breaching whale, when we encounter another kind of jump cut—the sudden appearance of Peter, our enthusiastic skipper, arrayed in a wool skipper's cap, yellow oilskins, deck boots, and accordion. He leaps onto the boat's doghouse in the middle of our carefully arranged shot of a tiny island and its colony of Arctic terns. Without introduction or apology, he pumps his accordion and belts out a Newfoundland sea chantey: *I'se the b'y that builds the boat and I'se the b'y that sails her.*

We are already familiar with a peculiar reality—that the presence of a movie camera triggers psychological jump cuts in most human beings. Easygoing folks freeze up. The articulate turn reticent. Experts get lost in thickets of knowledge. The well-spoken unloose streams of unpunctuated consciousness. Gentle souls charge the camera with violent soliloquies, some so forceful that even weeks later, when we're alone in the dark editing room as the reels of workprint and audiotape crank through the gears of the flatbed, we find ourselves glancing over our shoulders to see who the person on the little optical screen is yelling at.

Peter is performing his audition straight to camera, a filmmaking conceit we avoid. Unbeknownst to him, we are shooting what is known in the business as MOS, that is, film without sound (Hollywood legend claims that the acronym derives from a German director of old who could only pronounce the phrase "mit out sound"). Peter brays: *Sods and rinds to cover the flake! Cake and tea for supper! Codfish in the spring o' the year! Fried in maggoty butter!* We will never hear a word of it.

Meanwhile, the terns are provoked to panic, rising off their eggs in a shrill whirlwind. The effect is unflattering to our enthusiastic host, so we surreptitiously turn off our cameras.

Filming wildlife is as challenging as shadowing movie stars. The tricky part is to avoid becoming paparazzi. When we fail, our failures live on, emblazoned on celluloid in the form of our subjects' hind ends running, swimming, floating, and flying away with all the haste and by whatever conveyance they can muster. The backsides of movie stars may be salable subject matter. Not so for wild animals.

Since most creatures commit without hesitation to the entirely rational decision to flee our cameras, we soon own miles of jump cuts—a reject library of hundreds of individuals of countless species making the abrupt behavioral transition from normal to terrified, leaving behind the puzzled Zen of an empty frame.

So we cultivate cunning, hiding ourselves, waiting patiently, allowing our subjects to either forget about us or get to know us over the course of hours, days, or weeks. A few can't or won't habituate and prefer to challenge us—advancing or charging or threatening or snarling. Others circle the conflicted eddy between flight and fight until they're sucked down the drain of uncinematic stasis. Once in an extraordinary while we encounter the complete disregard we fervently desire.

On a sandy Newfoundland beach, late on an afternoon dulled by fog, we come upon innumerable wild animals gathering on the pebbly shore. They are willfully abandoning their natural realm and exposing themselves to predators of land and air. They are capelin, and they're finally spawning, an annual ritual known in these parts as the scull. Tens of thousands of fish are turning the waves into polished silver purses that roll ashore and spill their wriggling treasure onto the beach, where many dangers await them. Yet the urgency of their mission trumps fear.

The fish land in threesomes, two males enveloping a female, all lying on their sides while beating their tails, struggling to swim *up* the beach through the backwash. They are small fish, between five and seven inches long, and their pretty bicoloring of silver undersides and bottle-green backs—perfect camouflage in the deep—provides no disguise whatsoever on a beach of tawny pebbles, even in falling darkness. The capelin's efforts to make landfall are so single-minded and frantic, and so many of them are driven to it at the same time, that you can't help but imagine that the sea has been suddenly electrified, and the fish's only hope of survival is to escape it. Which, in an evolutionary sense, is true.

The male capelin, larger and more robust than the females, bear two pairs of spawning ridges, one pair each on their dorsal and ventral sides, designed to hold the female in place as they land, regardless of which position the male occupies in the threesome. The males have grown these spawning ridges in the past few weeks, producing them from the elongated scales for which the species is named *villosus* (Latin: hairy). Between one set of waves and the next, locked between the shaggy ridges of two males, the females shed their 6,000 to 12,000 eggs, and the males release their sperm. The sucking retreat of the sea buries the fertilized spawn in the pebbles of the beach, and the spent fish slip back with the outflow. The adult males, salmonlike, will die soon thereafter. Many females will survive to breed again, and those that do, larger and able to brood more eggs and stronger offspring, will likely be among the most productive and valuable contributors to the fish stock one year from now. Marine conservation biologists have dubbed valuable fish elders Big Old Fat Fecund Female Fish, or BOFFFFs, and argue that wise fisheries management would foster their protection.[1]

But not all capelin slip back offshore. Some are caught by five local children celebrating what might as well be called the feast day of Saint Capelin in Newfoundland. Wearing rubber boots and hand-me-down jeans, sagging wool sweaters wet to the elbows, they shout and squeal and charge into the oncoming waves. Reaching down with bare hands, they close the nets of their fingers on fistfuls of fish and toss them into metal pails. A black-splotched white cat walks daintily along the shore, tapping the fish with her fingertips, while a black-and-white dog gallops between the cat and the kids, barking and snapping at the water, snaring capelin between his teeth. He bears an uncanny resemblance to the extinct local breed known as the St. John's water dog, that floppy-eared, tuxedoed precursor of retrievers. Legend says those dogs were left by the Vikings when they abandoned Newfoundland. Whatever his true lineage, this canine—veering between the overflowing waves and the children's overflowing buckets—has rediscovered a version of Ægir's magically refilling tankards.

The sea heaves with the bodies of fish for whom safety and

opportunity lie in numbers—as it does for their predators. The dichotomy between benefit and risk, hunted and hunter, converges in one frenzied stream of action along the shore of the scull. Already the gannets have arrived in their characteristic frantic flocks, white birds by the hundreds shooting from sky to water in staggered, asynchronous dives, like jagged bolts of lightning on a pewter sea. They rise fifty or a hundred feet above the water, hover, fix their sights on the school, then tip over and fall, stroking hard once or twice with powerful wings before folding up and disappearing into the wet.

Atlantic puffins (*Fratercula arctica*; Red List: Least Concern 2008) bob high on the water, unperturbed by the spears of gannets shooting past them. They dive from the surface or from low in the air and drive themselves underwater on oarlike wings and rudderlike feet, flying swiftly into the heart of the fish school to snatch their bounty from the confusion. When they pop back to the surface, they are as buoyant as decoys and miraculously decorated with one or more silvery capelin draped across their gaudy bills, bedecked with horny nuptial plates in scarlet, yellow, and blue. Newfoundlanders call the puffins sea parrots for their outlandish beaks, which Arthur Cleveland Bent deemed "suggestive of the false nose of a masquerader." Even under the weight of the bill and fish, puffins fly fast and hard on tiny wings, powering seamlessly between water and air. Bent described their unique flight:

> I have often seen them emerge from a wave, fly across the trough and enter the next wave without apparent change in their method of propulsion. Again I have seen them come out of the water flying, only to plunge down into the water and continue the flight below the surface. On the surface they paddle along skillfully like little apoplectic short necked ducks and their small orange red legs are plainly visible.[2]

On the edges of the puffin flocks, common murres (*Uria aalge;* Red List: Least Concern 2008) raft by the hundreds, birds sculpted as starkly black and white as if lit by stage lights, even in the fog. They share a proclivity for facing into the wind, afloat on their white bellies, loonlike bills tipped into the air—a tableau of wind barbs on the weather map of the sea. Their hunting strategy involves diving the edges of the capelin school all the way to the bottom of it and attacking fish on the flight up—a method more strenuous than the puffins', yet infinitely easier than their own arduous work earlier in the breeding season.

Prior to the scull, in the spring months while the murre chicks are brooded and hatched, the adults await the capelin, who are waiting for the sea-surface temperatures of Newfoundland's shores to warm to at least 42 degrees Fahrenheit. The capelin bide their time in gender-segregated schools—the males inshore, the females offshore in a stratum of the Labrador Current known as the Cold Intermediate Layer. Massed by the millions in water hundreds of feet deep and colder than the freezing point of fresh water, the females slow their metabolic rate and enter a state akin to torpor. The chill acts as thermal camouflage, hiding them from their primary piscine predator, the northern cod (*Gadus morhua;* Red List: Vulnerable 1996), whose reflexes and drive are likewise slowed by the frigid temperatures.

But not so for the murres—warm-blooded, limber of muscle, speedy of thought, even in the bitter water. Somehow, in the course of the long evolutionary rally between predator and prey, common murres discovered the schools of female capelin assembled in the dark in the freezing cold on the bottom of the sea on the Grand Banks. Someway the murres trained themselves to perform the Olympian dives required to descend 500 feet and remain below for three and a half minutes in pursuit of difficult-to-reach fish trapped in a barrel of cold and black. By some means, the impossible was perfected—a perfection contributing in part to the extraordinary ecological success of this species, as measured by its abundance.

Twenty-one million breeding common murres inhabit the northern circumpolar crown of the deep blue home, along with uncountable millions of their young, who, en route to sexual maturity, as late as their eighth year, must learn the impossible for themselves.

Not all capelin come ashore to spawn, although this was probably the original breeding strategy of the species *Mallotus villosus*, which first arose in the North Pacific and later colonized the North Atlantic in the high-water periods between ice ages. The capelin of today retain beach-spawning populations only in the North Pacific and Newfoundland. The remaining stocks breed deep underwater off West Greenland, Iceland, and western Newfoundland; in the Barents Sea and the St. Lawrence River; and on the southeasternmost shoal of the Grand Banks.

Of all the underwater Atlantic breeding sites used by capelin today, only the Grand Banks was free of ice during the last glacial maximum 12,000 years ago. At that time, much of the water of the North Atlantic was locked up in massive icecaps, and from that frozen world the Grand Banks hove out of the sea, becoming a truly newfound land, a Grand Island, high and dry and ringed with beaches ideal for capelin spawning. Today's Grand Banks spawners are believed to be descended from those pioneer beach breeders, who, generation by generation, returned to natal shores, even as the icecaps melted and the island sank beneath the waves. Today's fish spawn on shoals between 100 and 200 feet deep.

So perhaps it's not an inconceivable jump cut that common murres fishing for capelin in shallow surface waters developed the extraordinary skill of fishing 500 feet deep among sluggish schools of female fish. Perhaps, generation by generation, the murres learned to extend their dives in accordance with rising sea levels, shifting beaches, the ebb and flow of predators, the changing ways of prey.

Everything arises, everything falls away.

Now, in 1984, with snow flurries melting onto the tongues of the waves, we stand in water to our knees, holding our cameras under as the pulse of the deep blue home drives a braided mass of capelin onto the shore and draws it back again. The capelin scull is transforming the beach under our feet. Roe, ripened inside cool female bodies during their tenure in the cold, is filling the interstices of sand with millions of lustrous amber globes little bigger than poppy seeds. Despite the overcast and the snow, these eggs gleam with the sheen unique to seaborne life—a mass stranding of glass fishing floats from a miniature world. They glisten among the multicolored pebbles of quartzite, sandstone, limestone, and shale, ground from the land by the waves. They sparkle beside the pulverized remains of Newfoundland's beach fossils, its trilobites and the earliest jellyfishlike animals.

The bounty ashore is so great that some birds abandon normalcy for novelty. In near darkness, upon the departure of the children and the dog but not the cat, two gannets fall from the sky to the beach, fold up their sails, and stretch their necks, bills swiveling right and left. They begin to waddle, ungainly bodies tilted backward, enormous wings flicking open for balance. They patrol the wave edges and proceed to stab capelin from the sand, shaking their catches hard, catlike, in the lances of their bills before tossing them to the air and catching them again, swallowing them headfirst and whole.

It's too dark to film now, even though the snow tumbling in huge slow-motion doilies is lit by a golden alpenglow and even though the crests of the waves coming ashore scintillate with gilded fish. In Newfoundland English, the hour is *duckish*. In filmmaking parlance, it's magic light, a breathtaking and ethereal idyll, the colors animate and shifting with life, saturating our eyes through to our blood.

A flock of tiny shorebirds cascades through calm air like a flurry of wind-whipped snow to land on the beach and skitter sideways, as softly as pink shadows chasing the waves. They stitch the roe from gravel to gizzard, sewing themselves to the fabric of the sea. It's not inconceivable that these tiny travelers fueled themselves on the bounty of Isla Rasa's *estero* a mere three months and a staggering 3,400 miles ago.

We sit on the beach as night steals away even the reflections of the visual world, leaving us immersed in a theater of sound: the hiss of withdrawing waves, the fluttering tails of fish thrashing the gravel, the whistle and splash of birds hurtling from air to water, the jump cuts from life to genesis to death.

20

Lament for the Thirty Million

DESPITE THE MIGRATORY BIRD TREATY of 1918, common murres and their close brethren, thick-billed murres (*Uria lomvia;* Red List: Least Concern 2008), are victims of an enormous wintertime hunt in the waters off Newfoundland and Labrador. The turr hunt, as it's known locally, is an outgrowth of an ancient subsistence kill by the Innu and Inuit peoples and of a nearly four-hundred-year-old subsistence kill by Newfoundland's European descendants. Today it's a purely modern affair and the only one of its kind in North America in which murres can be legally hunted for food or sport by nonaboriginal peoples armed with semiautomatic shotguns riding speedboats into the darkness of the winter waters in pursuit of migrating and overwintering birds.

As of 1984, up to 900,000 birds are being shot annually. Most are juveniles, including the flightless young who have already survived formidable odds, including the inconceivable leaps from their nesting ledges—in some cases, hundreds of feet above the water—at a time when they have not yet grown their primary feathers and cannot fly. Floating, tumbling against the cliffs, each youngster is accompanied on its free fall by a parent who descends closely alongside, braking its own speedier descent by swaying, Mary Poppins–like, on the four parasols of flared wings and webbed feet.

Many of the murres hunted by the people of Newfoundland and Labrador are flightless birds of the year, afloat on the water and paddling hard in the process of swimming the first six hundred miles of their winter migration. They are entirely helpless along this route and are tended and fed by their fathers (their mothers fly south), who also fall to the shotguns.

In 1993 the Canadian Wildlife Service will begin to control this slaughter, setting seasons and bag limits—ten birds per hunter per day over a three-and-a-half-month period. The new rules reduce the annual kill to about 300,000 a year (assuming the hunters abide by the rules), a number theoretically holding steady as of this writing.

Yet even as the turr hunt has declined, another killer has emerged in these waters, one far harder to rein in. Long-term surveys of dead and stranded birds washing ashore on Newfoundland's beaches reveal a growing epidemic of oil fouling. In 1984 just over 2 percent of beached murres showed signs of oiling. By 1997 the number had grown to nearly 77 percent, the highest rate in the world over a large region and growing at about 3 percent a year—even in the absence of a major oil spill.

Just as some of the deep blue home's greatest ocean currents converge in the waters around the island of Newfoundland, so do some of the most congested shipping routes, those between Europe and North America. Sadly, the most heavily oiled birds are found in the region of Cape Saint Mary's, the epicenter of currents, shipping, and birds. Chemical analysis reveals that 90 percent of this contamination comes from the heavy fuel oil driving large oceangoing vessels. The fouling is likely a result of ships illegally discharging oily bilge and ballast water at sea. The consequences are lethal to some 300,000 murres and dovekies (*Alle alle;* Red List: Least Concern 2008) off Newfoundland each winter.[1]

The second oil source in these waters is the enormous drilling field of the Grand Banks and its numerous rigs, including the largest in the world, the production platform *Hibernia,* storing more than a million barrels of crude oil and housing a 185-member crew. With this and two other drilling megaprojects, plus a new field under development, the oil industry is rapidly transforming the Grand Banks from a high-octane fishing ground that once powered civilizations on both sides of the Atlantic into a high-wire risk stretched between profit and hubris across some of the stormiest and most biologically productive waters on Earth.

Newfoundland's ice-laden seas have already claimed one victim of the oil industry, the semisubmersible drilling platform known as the *Ocean Ranger,* once the largest rig of its day. In February 1982, drilling the Hibernia Oil Field, the *Ocean Ranger* suffered a perfect convergence of 100-knot winds and 65-foot waves, which punched open a porthole in the ballast room 28 feet above sea level. Within five hours *Ocean Ranger* put out a mayday call, and within thirty-eight minutes of that distress call, its final message: *There will be no further radio communications from the* Ocean Ranger. *We are going to lifeboat stations.* In treacherous seas on a heavily listing rig, the launch and viability of lifeboats proved impossible. All eighty-four souls aboard perished as the biggest rig on Earth sank to the cold, dark bottom of the Grand Banks.

Drilling designs have improved in the aftermath of that disaster, yet accidents continue. In November 2004, 44,000 gallons of crude spilled from the *Terra Nova* megaproject. No one knows how much marine wildlife was killed, since little or no attempt was made to either clean up or follow up in a meaningful time period, a response as effective as a coverup. A full two years after the spill, a Canadian Wildlife Service report finally concluded that between 10,000 and 16,000 murres and dovekies within the *Terra Nova* spill zone were "put at risk"—a euphemism for killed.[2]

Even in the absence of spills, the Grand Banks oil industry increases the level of boat traffic in the region and therefore, in all likelihood, the number of smaller shipboard oil spills, intentional or otherwise. Dedicated fleets of shuttle tankers ply the hundreds of miles of waters between the rigs and the village of Come By Chance, an erstwhile fishing port transformed into Newfoundland's only oil refinery. The village is nestled deep inside Placentia Bay for safekeeping from the worst of Newfoundland's weather, though its snug placement ensures that every shuttle tanker to and from the Grand Banks oil fields passes directly abeam of Cape Saint Mary's.

Canada is the seventh-largest oil-producing nation on Earth, with reserves second only to Saudi Arabia's, generating more than three million barrels a day of oil, tar shale, and natural gas. But the real story here is that one million barrels a day are exported to Canada's insatiably oil-thirsty neighbor to the south. The United States sucks up virtually every drop of exported Canadian oil, and Canada is the United States' biggest supplier, making citizens on both sides of that American border more deadly than any turr hunter to the incomparable wildlife of Newfoundland.

In necropsies, the silence of the bodies of oiled birds speaks volumes of their suffering: of the hypothermia resulting from oiled feathers no longer able to insulate against the cold; of the malnutrition resulting from the hypothermia, which prevents the birds from diving deep enough or long enough to forage for food; of the deep anemia resulting from the shock and stress of hunger; of the poisoning from ingesting or inhaling oil during the extensive preening the birds perform at the first sensation of cold or wet.[3] Chronic oil fouling initiates an irreversible death spiral in a world of cold water and colder air. Most of Newfoundland's beached murres die in the winter, when thermal stress is at its annual peak, and when the added stressors of the turr hunt are underway.

More than 30 million seabirds reside in or migrate through Newfoundland waters each year. Many fall afoul of oil, either from diving through it or from eating those contaminated by it. The species affected include northern gannets, Atlantic puffins, black guillemots, dovekies, razorbills, several species of eider ducks, long-tailed ducks, northern fulmars, black-legged kittiwakes, great black-backed gulls, herring gulls, glaucous gulls, and Iceland gulls.

The ancient sanctuary of Cape Saint Mary's, chosen by millennia of seabirds for its incomparable natural refuge, offers no protection from these modern troubles. Nor apparently do modern solutions. A meta-analysis by Francis Wiese of Memorial University and Pierre Ryan of the Canadian Wildlife Service, found that even as technology and legislation addressing marine pollution improve, the oiling rates of seabirds continue to rise.[4]

Cape Saint Mary's is not the only vulnerable location along the narrow fingers of land and deep fjords of water marking the island of Newfoundland. Baccalieu Island, the northeasternmost point of the Avalon Peninsula and therefore the northeasternmost point in North America, apart from Greenland, comprises two square miles of cliffs and glacier-stripped coastal heathlands. Baccalieu is a mostly treeless realm shorn by the wind into low barrens soggy with mosses. The few balsam-fir trees that subsist in this wet world form stunted forests known locally as tuckamore.

Yet as much as climate and soils have engineered this (or any) ecoregion, Baccalieu's birds have shaped it too. The island is home to the world's largest colony of Leach's storm-petrels (*Oceanodroma leucorhoa;* Red List: Least Concern 2008)—close relatives of the least storm-petrels nesting in the talus slopes on Mexico's Isla Partida. More than 7 million Leach's storm-petrels nest in burrows in the moss and among the roots of the tuckamore on the sparse real estate of Baccalieu Island.

These are diminutive birds, smaller than robins, who collectively alter the recipe of this ecosystem in powerful ways: overturning the soil, binding it with guano fertilizer, stirring in feathers and carcasses, enriching it with their fishy stomach oil, seasoning it with salt excreted from their perpetually runny nostrils. Storm-petrels arrive at the island in the millions every evening of every year between March and September and depart in the millions every morning. Their commute to and from the deep blue home blurs the borders between land and sea.

Before they were protected under Canadian law, Leach's storm-petrels were hunted with bait and used as bait by Newfoundland fishermen, who trolled codfish livers in the water and snapped homemade bullwhips into the throngs butterflying over the oily sea—an effective if cruel technique that slew birds by the hundreds. Locals stole fattened petrel chicks from their nest burrows, threaded wicks down their gullets to draw up the stomach oil, and burned them as candles.

Colonies of storm-petrels once graced many islands from Maine to Labrador, but most disappeared as cats, dogs, rats, pigs, foxes, mink, skunks, and sheep were introduced. Arthur Cleveland Bent observed a large and populous colony of Leach's storm-petrels on Seal Island off Nova Scotia in 1914, only to find it nearly obliterated a few years later—victim, apparently, of the lighthouse keeper's Newfoundland dog, who hunted petrels obsessively throughout the breeding season.

Experiments from the 1980s with nesting Leach's storm-petrels in a surviving colony, Grand Island off southern Newfoundland, found that birds exposed to crude oil or to the oil-dispersant emulsions used in cleanup efforts lost more eggs and chicks than did control birds—even though the level of oil exposure was sublethal to adult birds and even if only one adult of the pair was oiled. Most of the birds doused externally with a medium dose of oil—in this case, 0.5 milliliter of Prudhoe Bay crude—abandoned their eggs the same day, presumably so they could fly to sea and preen. Those receiving lesser doses tended to incubate their egg—and therefore unwittingly oil the egg—only to abandon it that night, even though no partner had come to relieve them at the nest.[5]

Parent birds oiled after their chicks hatched tended to brood their chicks—and thereby oil them—for a day, then abandon them that night. Many of these orphans bore the classic signs of rejection: pecking wounds on heads, backs, legs, and bellies, presumably inflicted by one or both parents or by a neighbor bird or birds.

Breeding success for adults conscripted into this study generally returned to normal the following year—except in the case of birds administered the highest doses (1.5 milliliters) of either Prudhoe Bay crude or Corexit 9527, the dispersant designed to break an oil slick into small droplets, presumably for easier diffusion into the environment. Many fewer of the storm-petrels contaminated with higher levels of the oil or emulsant returned to Grand Island to breed the following year, intimating that their part in the experiment had proved lethal after all.

21

All Time Is Now

I N HIS CROWDED LABORATORY at the Village Inn in Trinity, Newfoundland, Peter shows us his buckets. Pails full of water are connected through an alarming cobweb of wiring to each other and to a veritable Radio Shack of hand-built measuring devices. Peter is researching humpback whale communication. The central question in his mind is how animals such as whales lead such seemingly complex lives, using, in his words, "so few signals, signs, or symbols." Although there's an explosion of research going on in 1984 in the science of marine bioacoustics, fueled largely by investigations into the intensely songful lives of humpback whales, Peter seems unconcerned with those inquiries.

Instead, his buckets are full of barnacles. His supposition is that whales use the barnacles hitchhiking on their skin as binary on-off switches to send messages to other whales. As of 1984, his hypothesis is under development. Twenty-six years later it has matured into a self-published online dissertation involving what he calls Rhythm Based Communication: "the biological rhythms shared between two organisms so that synchronization occurs."

After synchronization, Rhythm Based Communication is made possible by an organism's perception of lateness relative to on timeness. Organisms, through synchronization, arrive at a common rhythm, and within this synchronized rhythm, transmit and receive messages using combinations of ON-TIME, LATE, OFF-TIME and EARLY messages. Such information flow is Rhythm Based Communication.[1]

Peter is a compact man with the tightly rolled and unpredictable energy of ball lightning, that controversial phenomenon bouncing through the conventional world with preternatural brightness. If Tinker Bell were grown-up, Canadian, male, and human, she might resemble Peter, who is also quirky, funny, narcissistic, sulky, tempting, mercurial, sweet, and indirectly menacing. Much of the time he too can hold on to only one feeling at a time, though he equalizes this temperamental shortfall with an abundance of feverishly contagious enthusiasm, his own private fairy dust.

Peter's light shines brightest when he's moving fastest: full steam ahead aboard *Ceres* in the fog; dodging moose in the snow in his old station wagon. *I won't die if you just believe.* He mixes Manhattans in his pub at the Village Inn with a joyful flourish of *three secret ingredients.* He sings and plays piano and accordion, and although he is not a native Newfoundlander, he has studied the part and enlivened the staid character of the hard-worn islander with his own stagy enthusiasm.

Underlying all his theories about barnacles and whales is a fascination with time, the "enormous enigma," he says, encompassing the future in ways so vast and impenetrable—except, possibly, by the human mind—that it deserves its own bigger and better name. He calls the real thing TIME. To get to TIME, he unwinds balls of string down convoluted mathways, squeezes under philosophical gates, hurtles over hidden psychology:

One's mind has a 'Mental Thought Process, MTP,' (of one's 'conscious mind'), which can 'transduce' incoming signals in 'Conventional time Ct or t' (future-to-past) information, into 'RhythmicTime RT or T' (coded 'informaTion'), which is the suggested Quantum Mechanical form of active thought processes ('MTPs'). Mind can also use a system of 'MEmory Detection And Retrieval,' ('Medar,' pronounced 'Maa-dar') to recall past events, so that they can become one's 'Now TIME.'

At the end he arrives at the simplest and most Zenlike of realizations: all time is now.

It's codfish weather, as they say in Newfoundland, foggy and cold. Between patches of invisibility near the village of Bareneed, we come upon a wooden fishing boat and three fishermen retrieving a trap in shallow water. They are father and adult sons, we imagine, a working unit as old as fishing time. The boat is not unlike a Mexican *panga*, about twenty-five feet in length, open-decked and broad in the beam, designed to accommodate a sizable catch. The boat's single concession to Newfoundland's foul weather is a small wheelhouse, about the size of an outhouse, set far aft, enclosed on three sides, open to the stern. It doesn't offer much break from wind and waves, just enough for three men huddled inside, peering through a windshield as narrow as a squint, to work their way home.

These fishers possess no radar or LORAN (Long Range Aid to Navigation), and it's years before the Global Positioning System will be widely available. What these men do have is knowledge accumulated and bestowed upon them by their forebears. They also have, tethered off their stern, a weather-beaten rowing dory of the classic Grand Banks design, flat-bottomed, high-sided, pointy in the bow and stern, flaring upward in pretty curves. Only a generation ago, maybe two, Newfoundland men rowed these little ladies of the North Atlantic out on the Grand

Banks to seine for cod. Now they're mostly relegated to inshore tenders.

The men's cod trap is another design endemic to Newfoundland: a net formed into a box, with four sides and the bottom enclosed and the top left open at the surface. The effect is to produce a submerged room, maybe fifteen feet on each side, whose geometry is maintained by four buoys bobbing at the surface and four anchors sunk to the bottom. The secret of the trap involves an underwater net fence, known as a leader, made fast to the shore and running out to a small doorway in one wall of the trap. Cod swimming in shallow water come upon the leader and try to get around it by swimming to deeper water, only to be led into the trap. The design is virtually unchanged from the first cod trap built by the Newfoundlander William Whiteley in the 1860s, which liberated the many men once needed to attend seine nets and allowed a family to work a trap, or even one man alone.

The three fishers off Bareneed, born and bred on cod traps, are pulling their trap now, nets in hand, untangling a few meager codfish. These actions are second nature to them, leaving them free to investigate the unlikely appearance from out of the fog of our newfangled inflatable Zodiac. The youngest man, in red baseball cap and yellow oilskins, smiles hesitantly. The second young man, in black baseball cap and black oilskins, ducks his head and keeps it down, stealing glances sideways. The older man, in a yellow wool flat hat straight out of the Irish Sea, stands tall and turns to face us, unblinking and suspicious, though not hostile.

Seen any whales? Peter calls across the water.

The older man shakes his head.

We're looking for whales, shouts Peter.

Nothing.

The two younger men straighten up and square around to stare too.

They're late this year, explains Peter.

Only the barnacles are answering.

The older man shrugs and turns back to the net in hand.

Humpback whales are recovering from the first sixty-six years of the twentieth century, when upward of 200,000 of their kind —an estimated 90 percent of the global population—were killed by commercial hunters. The whaling moratorium imposed in 1966 is beginning to reap dividends in 1984, and the original 700 survivors in the North Atlantic are breeding their way toward a whisper of sustainability.

In Newfoundland this means humpbacks are once again chasing the capelin scull. Which means the whales are hunting alongside cod, who are likewise tailing capelin, so whales are running into the cod traps too. Newfoundland's fishers do not wish to catch whales in their traps but are nonetheless snaring a growing number of behemoths, who do terrible things to the gear, which keeps them pegged to the surface and unable to dive, eventually starving them to death.

I can get a whale free from your gear, shouts Peter. If you ever need that.

Now he has their interest. All three men stop for a harder look. Is there someone crazy enough to do that?

Peter smiles his infectious smile.

Just call over to the Village Inn in Trinity, he says. We'll rescue your gear.

Gunning the outboard, he pivots the Zodiac, kicking up a roostertail of spray, and dashes off into the fog toward the next cod trap, somewhere up the coast. In flight, he cheerfully bellows a sorrowful ballad:

> *Poor fishermen, been out all the day;*
> *Come home in the evening full sail up the bay,*
> *There's Kate in the corner with a wink and a nod,*
> *Saying, Jimmy or Johnny, have you got any cod?*
> *And it's hard, hard times.*

• • •

In 1984 the protection of Newfoundland's breeding seabird colonies and the reappearance of humpback whales is converging into a circulating stream of bitterness, mostly among those on this island who see regulations as an attack on their hunting, fishing, and gathering way of life. There is a genuine link between the protection of wilderness and the success of the Newfoundland way of life. But it's a link with a long leader, so to speak, forcing a fisherman far offshore when he wants to stay inshore—even though the inshore of his unfettered tradition will certainly kill his way of life, whereas the regulated offshore just might save it. Nevertheless, many fishers cannot see to salvation through the thicket of their dwindling livelihoods. And salvation, anyway, requires TIME, that enormous enigma, which few, whether fishermen or fisheries managers, have the patience for.

One of the newest worries keeping the pubs afloat in every outport of Newfoundland's remote fishing villages is the birth of a commercial fishery for capelin. In 1984 the market for roe has just arrived from a sushi-hungry Japan. Suddenly there's money to be made from selling the children's catch—or rather from sending out boats to catch capelin before they run inshore to the beaches where the children will catch them. Yet even those Newfoundlanders involved in the capelin fishery are concerned, since capelin roe, obviously, fuels the next generation of capelin, and capelin fuel cod, Newfoundland's currency.

Qualms about capelin invariably lead to fears about the inconceivable: the decline in the cod fishery that has sustained this island for four hundred years. As early as 1550, 128 cod-fishing boats sailed to Newfoundland from Europe every summer. By the late seventeenth century, the catch approached 100,000 metric tons a year. It doubled a hundred years later and doubled again in the nineteenth century to 400,000 tons a year.

But now, as Peter delivers us to another cod-fishing boat, we see how little is left. Two old men in threadbare oilskins and handknit woolen gloves worn through at the palms are pulling a trap mostly full of emptiness, seasoned with paucity—a half-dozen cod, and small cod at that, two feet or less. Worse, just offshore of their trap, only a gunshot away, as they say here, lies a much bigger vessel, a seiner, hauling in a bloated purse of capelin. The old men work with their backs to it.

Hallo, calls Peter.

They ignore us too. Still, we can't shake the feeling that the partially sunk waterlogged planks have been baited, the booby-trap set.

Atlantic cod grow to lengths of more than six feet and weights of more than 200 pounds. For centuries big fish and big catches were expected and delivered in these parts. Newfoundlanders landed an estimated 300,000 tons of cod a year until the 1950s and more during the 1960s, when factory trawlers—including foreign vessels fishing the Grand Banks outside Canada's territorial limit—expanded the fishery exponentially. In 1968 the annual catch exploded to an all-time high of 800,000 tons. But it was a Ponzi scheme, stealing from the future to pay for an illusion of present-time prosperity.

Motoring from cod trap to cod trap, we see clearly that something terrible is happening. The fish are harder to catch and smaller, a tragically classic measure of unsustainability. It's also a sign of the novel way the cod themselves are struggling to survive: by adopting a last-chance strategy of breeding younger. The median age of sexual maturity for female cod in 1959 was just over six years and for males about five and a half years. By 1979 it had fallen to less than three years for both sexes.

In 1984 the fishers of Newfoundland are worried. But not enough to stop cod fishing, since there's nothing else to do here. The Canadian government maintains a sunny optimism about the ability to scientifically manage this fogbound fishery. Yet their management plan will prove fouled by bogus data on two fronts: chronic overassessments of the size of the cod stock and chronic underreporting by fishermen of the size of their catches. Everyone in these parts, as they say in these parts, is *scrimshank:* hesitating and avoiding the issue.

The fisheries managers also manage to ignore the grim lessons of the twentieth century: two invincibly enormous fisheries have already collapsed from combinations of overfishing and climate change. The Peruvian anchovetta, once the largest fishery on Earth, with catches above 10 million tons in the late 1960s, collapsed completely in the 1972 El Niño. Thirty years earlier, the California sardine fishery, made famous by John Steinbeck's *Cannery Row,* collapsed from overfishing during a period of cooling. In 1984 history's most enduring lesson—how little we learn from history—is repeating itself in Newfoundland.

Seen any whales? Peter tries again, calling to the two old men in their battered dory.

Aren't any yoi, says one man without looking up. No cod, neither.

Well, I can get a whale free from your gear, shouts Peter. If you ever need that.

They look up but don't answer, their faces creased with weathery skepticism.

Nevertheless, we know. At this point, a whale in their gear will drown them too.

22

Trophic Cascade

O N ANY GIVEN DAY, some 4 million commercial fishing vessels ply the deep blue home. Until recently, it was a glorious business. The growth of seagoing technologies paralleled the growth in the annual global fish catch. Fishing got easier and catches grew bigger.

But 2000 marked a turning point. The global catch, which had grown 600 percent between 1950 and 1997, peaked at 96 million tons, despite better technologies and intensified efforts. Since then it has fallen by more than 2 percent a year and maybe more, based on inaccurate reporting from China and Peru.

Paradoxically, modern industrialized fishing has become so efficient that it is now supremely inefficient. One of the worst culprits is a form of fishing known as longlining, whereby a single boat sets monofilament line across sixty or more miles of ocean, baited with up to 10,000 hooks. An estimated 2 billion longline hooks are set worldwide every year, primarily for tuna and swordfish, though longliners inadvertently kill far more who take the bait, including some 40,000 sea turtles, 300,000

seabirds, and millions of sharks annually. These unwanted species, known coldly as bycatch, are thrown, dead or dying, back
into the ocean, where they amount to some 25 percent of the total global catch, about 88 billion pounds of life a year caught, destroyed, and discarded.

All told, pelagic longlines are the most widely used fishing
gear on Earth and are deployed in all the oceans except the circumpolar seas. But whereas they once caught ten fish per one
hundred hooks set, today they are lucky to catch one—an indica-
tor of the growing impoverishment of the deep blue home. Abet-
ting the destructiveness of longlining is another modern fishery
known as bottom trawling, which drags nets across every square
inch of the bottom of all the continental shelves on average once
every two years, with some regions trawled many times a season.
Razing vital sea-floor ecosystems is the brutal equivalent of fishing with bulldozers, leveling an area 150 times larger than the total area of forests clear-cut on land each year.

Added to longlines and trawlers is the technology of drift nets,
the nearly invisible curtains of monofilament blindsiding the life
of the ocean. Fishers in the North Atlantic set nets up to 150 miles
long and 1,600 feet deep in pursuit of sharks and monkfish. Left
untended, these nets set sail to randomly ensnare marine life. In
the course of operations, particularly in stormy seas, nets are lost
or abandoned, though they continue to fill with prey, which at-
tracts predators, who likewise become trapped, die, and decay,
attracting more predators. Composed of nonbiodegradable syn-
thetics, these deepwater ghost nets continue to fish with night-
marish efficiency for years.

The full consequences of modern fishing methods were unknown prior to the release in 2003 of a seminal study by Bo-
ris Worm and Ransom Myers of Dalhousie University in Nova
Scotia.[1] Their analysis concluded that, in a manner akin to viru-

lent pathogens, industrialized fisheries typically reduce the biomass of large fish by 80 percent within the first fifteen years of exploitation. In the wake of decades of such onslaughts, only 10 percent of the historical populations of all large fish—tuna, swordfish, marlin, halibut, skates, and flounder—are left anywhere in the ocean. Worm and Myers's analysis was based on the data that modern fisheries managers generally ignore—in this case, on catch reports from Japanese longliners of the 1950s.

Myers and Worm concluded that no one remembers how many big fish used to inhabit the sea or how big they actually got—a phenomenon of forgetfulness known as a shifting baseline. The few blue marlin left today reach only one fifth the weight they once did. But who remembers? A study analyzing 150 years of restaurant menus from cities such as Boston and San Francisco found, among many startling revelations, that lobster was so abundant in the nineteenth century that middle-class Americans snubbed it as food for the poor.

Many marine species are now under such intense fishing pressure that they don't reach sexual maturity and never even have the chance to spawn before they're caught. Market economics relentlessly drives commercially desirable species toward extinction. The endgame resembles a modern plague, a Black Death of the deep blue home, wherein the rarer and more endangered a species, the more money it generates, and the more people are willing to spend money and time in pursuit of it, hastening its demise. A single large bluefin tuna may command $150,000 on the Tokyo fish market. What Santiago isn't going to fight to land that?

A scientific review of the effects of human predation on wild species, ranging from cod to caribou, calculated that changes in body size and growth rate among our prey species outpace those changes among nonprey species by more than 300 percent.[2] Alarmingly, those animals that we hunt and fish show some of the most abrupt trait changes ever observed in wild populations. Unlike other predators, we target high numbers of large, sexually mature adults. The result is an unnatural selection for smaller animals that breed younger, driving some species, including cod, to become too small to eat their customary prey.

Newfoundland's cod currency began to decline dramatically in the 1970s after European trawlers recklessly caught prespawning, spawning, and postspawning fish, eventually leading to the collapse of both inshore and offshore cod stocks. By 1993 the median age of sexual maturity for northwest Atlantic cod had fallen from its known high of six years and its 1979 level of three years to less than two years—as close a statistical marker of desperation as numbers can describe. Fisheries and Oceans Canada at long last became alarmed enough to close the northern cod fishery for a two-year moratorium.

In 2003 Myers and Worm included the Atlantic cod on their infamous list of commercial species now reduced to 10 percent or lower of their historical abundance—perhaps as little as 5 percent in the case of Newfoundland cod. At the time of this writing, Canada's two-year moratorium has dragged into its eighteenth year, disconnecting 40,000 Canadians from a $500-million-a-year industry. A song by the Canadian Shelley Posen about the Newfoundland fishery provides as close a marker of desperation as music can describe:

> *Out along the harbour reach*
> *Boats stand dried up on the beach*
> *Ghost-like in the early dawn*
> *Empty now the fish are gone*
> *What will become of people now*
> *Trying to build a life somehow*
> *Hard hard times are back again*
> *No more fish, no fishermen.*

Most alarming, the cod crash appears to have restructured the food web into something unable to support cod anymore. At the same time cod were being overfished, so were other top ground-fish predators, including haddock, hake, pollock, cusk, redfish, plaice, flounder, and skates. Suddenly the cod-dominated ecology of the northwest Atlantic was absent its cod and also any other ground fish who might have swum in to take their place.

Small pelagic fish and invertebrate predators, primarily northern snow crab (*Chionoecetes opilio;* Red List: Not Evaluated) and northern shrimp (*Pandalus borealis;* Red List: Not Evaluated), once cod prey, were suddenly liberated from their predators. Predictably, their numbers grew, and with them a reciprocal response among *their* prey—the large herbivorous zooplankton, which declined by 45 percent. Meanwhile phytoplankton increased, presumably because their natural predators, the large herbivorous zooplankton, were declining. Lastly, nitrate concentrations, one of the most important factors limiting marine productivity, decreased as blooms of phytoplankton used them up.

The end result was the near mythical rearrangement of an ecosystem, a phenomenon never before observed and scientifically described in a marine environment the size of a continental shelf.[3] This reshuffling of populations and nutrients, which is known as a trophic cascade, is defined as the appearance of conspicuous indirect effects two or more trophic links removed from the primary effect. In Newfoundland, the loss of cod triggered an unprecedented four-trophic-level cascade: from crab to zooplankton to phytoplankton to nitrates.

The final verse of Posen's song concludes:

> There's some that say things aren't so black
> They say the fish will all come back
> Who'll be here to catch them then
> No more fish, no fishermen.

No one to remember.

23

Bone Rafters

THE VIKINGS ARRIVED in North America during the Medieval Warm Period, the luxuriant follow-up to the frigid centuries of the Dark Ages Cold Period, when European civilization slowed to a stutter. The Norse sailed the warming North Atlantic aboard clinker-built vessels known as *knorr,* boats with pine and oak planks overlapping like clapboards. These were not the sexy *skeid* of the Viking raiders, those dragonships reaching 100 feet in length and only 12 feet in width, manned by eighty men rowing at top speeds of 15 knots. The *knorr* were tubby, broad-in-the-beam cargo carriers, 50 feet long and 16 feet wide, open-decked, rigged with oars and sail, designed to transport a family, their slaves, their dogs, a couple of Icelandic ponies, a couple of cows, sheep and goats, their tools, farm goods, and housewares across open ocean.[1]

Knorr were far more seaworthy vessels than *skeid,* though the *Islendingasögur* (Sagas of Icelanders) report how often they were swept off course by gales or kept offshore by winds, sometimes for months at a time. The Viking adventuress Gudrid the Far-Traveler was shipwrecked aboard a *knorr* with her first husband

and spent a frightening, storm-tossed summer lost at sea aboard another *knorr* with her second husband, finally arriving in Vinland aboard a third *knorr* with her third husband, Thorfinn Karlsefni. The two succeeded not only in inhabiting the New World for three years but in parenting the first European child born there as well.[2]

As the Arctic scholar Robert McGhee points out, these Viking families were following seaways already pioneered and perhaps described to them by Irish hermit-monks—men who sailed aboard hide-covered boats known as currachs to the Faroe Islands and Iceland, perhaps as far as Greenland or even Newfoundland, maybe as early as the sixth century. The Irish seafaring abbot Saint Brendan the Navigator reported passing towering crystals of ice that rose to the sky.

Asail during the formidable Dark Ages Cold Period, these Irishmen, though brave and full of buoyant faith, were nevertheless, McGhee suggests, following pathways prescribed by the migratory swans, geese, ducks, and shorebirds who swarmed through Eire every spring en route to unknown islands to the north, returning every autumn with fledglings in tow as proof of a terrestrial world beyond. The seabirds, in turn, followed the circulating liquid trails laid down by fish, currents, and winds, the topography of the deep blue home.

The Vikings sailed their *knorr* to the greenest part of Greenland—the unpeopled southwest—around the year 1000. They may not have known it initially, but they were sharing Greenland—albeit at opposite ends of the island—with a tall, Eskimo-like group that archaeologists call the Dorset culture. These were true people of the ice, living a thousand frozen miles north of the Vikings' Greenland settlements, hunting from landfast ice and sea ice for seals, walruses, and whales. They used no boats but rode on inflated sealskins, a remarkable way of life that awarded them cultural dominance over the North American Arctic for at least 1,500 years.

The only word from the Medieval Old Norse of Greenland to survive to the present day is the name the Vikings gave to these people: *Skraeling* (perhaps from Old Norse *skrá:* skin, for their animal-skin clothing)—a word now meaning *barbarian* in modern Icelandic. There is no surviving word from the Dorset-culture people for themselves, because sometime during the two or three centuries that their Greenland tenure overlapped the Norse, the Dorset people were killed or assimilated by advancing Inuit people.

Like the Vikings, the Inuit capitalized on the Medieval Warm Period to expand their range, in this case eastward from Alaska, aboard *knorr*like vessels known as *umiak*. These were tubby, open-decked boats of lashed driftwood covered with the skins of bearded seals. They were named for their women rowers—as opposed to *qajaq*, or kayaks, the men's boats—and designed to carry a family, their dogs, tools, and household goods to new settlements.

Helping or hindering the Inuit on their way was Sedna, goddess of the sea, once a beautiful girl who persistently annoyed her father by refusing to marry. Sedna turned down one hunter after another, all the while brushing her hair and admiring her reflection in the ocean. Finally her father persuaded her to accept a particularly well dressed and seemingly prosperous hunter with a hidden face, who took her to his home far away on the sea cliffs, stripped off his furs, and revealed himself to be Raven.

Sedna was appalled. And helpless. She was forced to eat raw fish and to live in no home at all but a rough mat of fur and feathers. In her anguish she began to keen, a sound that grew loud enough to be heard above even the fiercest of Arctic winds, loud enough for her father to hear and feel pity. For days he paddled his *qajaq* toward her wailing, eventually rescuing her when Raven was away. But the bird soon found out and swept down upon the boat. When the father tried to beat him off with the paddle, Raven, in his fury, fanned his wings to make a terrible wind, driving the sea into monstrous waves.

Fearful for his life, the father threw Sedna overboard and told Raven he could have her. When the girl tried to climb back in, the father beat at her with his paddle until the frozen fingers of one of her hands fell off and sank to the bottom of the sea, becoming the first fat seals. Sedna tried again, swimming to her father's boat, grasping at it with her other hand. But again he beat her frozen fingers until these too snapped off and sank into the abyss, becoming the first whales and walruses. Despondent, Sedna allowed herself to sink too. On the ocean floor, surrounded by her seals, whales, and walruses, she became a goddess forever angry at men, appeased only at times by shamans who dove down to comb her black locks.

The Inuit moving east soon encountered the Dorset people, whom they named Tuniit and described as placid giants, easily driven away. In fact the Tuniit were already suffering a serious population decline at the time of the Medieval Warm Period, their numbers plummeting even as their culture reached its zenith in an explosive flourish of wood, ivory, bone, and antler carvings, including unmistakable images of long-faced, big-nosed Vikings. These last exquisite renderings may have been prayers for the return of the disappearing ice that was opening Arctic doors to Vikings and Inuit as it closed them on the Tuniit.

Sometime around 1400 the Dorset culture disappeared from the archaeological record—though the Inuit believed the Sadlermiut (meaning *First Inhabitants* in Inuktitut) living on remote islands in northern Hudson Bay were the last of the Tuniit. If so, then seventy of them persevered on these frozen outposts until 1902, when an American whaling ship anchored nearby and a sick crewman offloaded an unknown fever. Within a few weeks all the surviving Sadlermiut or Tuniit or Skraelings or whoever they were in their own minds fell ill and perished. *So large they broke the backs of bears,* wrote the Canadian poet Al Purdy in "Lament for the Dorsets," *so small they lurk behind bone rafters / in the brain of modern hunters.*[3]

We encounter the peculiar handiwork of modern hunters in drenching sleet on a desolate Newfoundland shoreline. We are *bivering,* as they say in the local dialect, shivering with cold as we strip out of parkas, sweaters, jeans, and long johns to wriggle into wetsuits. The sky wafts among us, ghosts of icy mist that cling, pass through, and drape over, separating us from each other, as if we've drifted into different universes. Some of these fingers are cold enough that we can hear the crackling of ice crystals tapping at our neoprene hoods.

As quickly as we can, we lurch toward the water, stumbling not just from the cold and the dark but from the phantoms here. This small beach composed of rocks and pebbles is now obscured by bones—thousands and thousands of rostra, scapulae, ribs, phalanges, vertebrae, crania, and mandibles—the remains of some of the 54,000 long-finned pilot whales (*Globicephala melas;* Red List: Data Deficient 2008) killed in the Newfoundland drive fishery between 1947 and 1971. We struggle to walk on such uneven pavement, slick with ice and green with algae, and are grateful to flop forward at the water's edge . . . only to find the horror magnified underwater, as our faceplates skim bare inches above the sea floor.

The beach may be overflowing with disassembled skeletons, yet it's only a veneer compared to what lies beneath the waves, as if a cargo container holding all the bones of the world had broken apart offshore and washed up here. I've forgotten my wetsuit gloves, and I reach now with clumsy frozen hands. These last surviving pieces of the whales have been rolled by the waves, their edges tumbled and polished into homogeneity, the sea mixing and matching once separate animals: males with females, calves with adults, relatives with rivals. Winter waves have broken ribs into fragments nearly indistinguishable from flipper bones. Summer storms have rounded spinous processes of vertebrae. Grounded icebergs, long ago melted, have scoured enduring tracks in the bottom, crushing bones to rubble and rubble to dust.

The water is 44 degrees, as cold as a cod's nose, according to Peter, who does not join us in its shocking embrace. The water is clear enough to award us an unhindered picture of the sloping seabed and its mountains of skeletons. We swim to the left and to the right and offshore and never find an end to the bones. It's a mausoleum of marble, utterly silent, except for the radio static of snapping shrimp, a eulogy broadcast from the past.

Yet life returns. Movements stir the dead and startle us: the slow-motion motions of northern sea stars (*Asterias vulgaris;* Red List: Not Evaluated), slower than sloths on their five arms. They are clambering among the bones, seeking food. Where one finds it, others gather, stacking atop one another in harmonious shades of yellow, pink, purple, orange—though there is nothing congenial about these piles. Each sea star, trying to elbow its way to the bottom, is hoping to make contact with the victim, a blue mussel or an oyster, and then to evert its stomach and begin to digest the flesh. The aggression is slow but fierce, and those stars that can't elbow their way in with all or most of their five arms will lose out on what may be one of only a few meals in this fleeting boreal summer.

Strange sentries abound among the bones, some so bonelike they appear to *be* bones, invisible at first and then unmistakable: squat, homely fish with hornlike spines around bulbous eyes, broad heads tapering to folded tails, backs prickly with spines. They are shorthorn sculpins (*Myoxocephalus scorpius;* Red List: Not Evaluated), whom the Newfoundlanders call horny whores. Their mouths are wide, down-set, thick-lipped, grumpy.

Yet ugliness is part of the sculpins' superb design, allowing these fish to open their mouths wide enough to swallow anything and everything that wanders by—crabs, capelin, sea stars, amphipods, marine worms—including prey nearly their own size. Stealth is their hook, followed by speed. When I swim too close, a sculpin grunts and jumps off its perch, pectoral fins folded tight to its body, and lands on a nearby bone. When I keep coming, it sashays resolutely away, undulating to the left and the right with each batlike flap of its wings.

Foot-long sea cucumbers (*Cucumaria frondosa;* Red List: Not Evaluated) mix with the bones too. These cylindrical, elongated creatures are white and bonelike themselves, except for their ten feeding tentacles, circlets of lacy branches surrounding their mouths and reaching into the water like a network of free-floating blood vessels. Each tentacle is coated with a sticky mucus designed to catch detritus and microscopic organisms afloat in the bloodstream of the current. When full, a tentacle folds into the mouth to be scraped clean.

It's a simple world, characteristic of the high latitudes, where species become fewer and ecosystems more concise. But it's also haunted by what used to be and no longer is—by the 60,000 long-finned pilot whales who once summered off Newfoundland.

The demise of that enormous community of life at the hands of local fishermen was born in the hungry years after World War II, part of a small-whale fishery initially targeted at minkes—though the Newfoundlanders soon found it easier to catch large numbers of gregarious pilot whales than the solitary minkes. And so every summer for twenty-four years the whales were herded by fleets of small boats toward beaches like this, where deep waters slope gradually to land. The gentle topography allowed for the most violent of hunts, as the men drove the whales shoreside, gaffed them alive into shallow water, then slaughtered them with lances.

As early as 1954, the bloody hunt was a source not of human food but of mink food, fed to caged animals raised for their fur. The nadir of the pilot-whale hunt was plumbed in 1956, when nearly 10,000 animals were killed on Newfoundland's shores. Thereafter the whales declined, and the hunt with it, until only 100 or 200 were caught in the final seasons before a halt was called in 1972.

Now, in 1984, in freezing rain only twelve years past the massacres, many of the bones have been softened by the waves. Yet others bear the marks of lance blades, sharp flakes knapped from the cervical vertebrae as if *jötnar* had once chipped their spearpoints here. The sense of tragedy has not washed away either, any more than it has, or ever will, at other battlefields like Gettysburg or Verdun, where the caliber of grief amid the mountains of bones continues to wound visitors even long afterward.

We flop ashore, exhausted, our fingers and lips blue. The sleet has turned to snow, feathers of white drifting onto a white beach. The bones, brittle with ice, crackle beneath our feet.

24

Soundsabers

NEWFOUNDLANDERS call long-finned pilot whales pot-
heads, for their bowling-ball foreheads. In fact, these melons, as they're officially known, are fat-filled organs critical to the
transmission and reception of sound underwater, and all odontocetes (Greek *odontos:* tooth; Latin *cetus:* whale) have them.

The toothed whales and dolphins echolocate as a way to navi-
gate through the underwater world, where sight is severely limited
by the density of water and where most serious hunting is done
in the dark of night. Like bats, the toothed whales see through
sound, and the melon is the eyeball of their ears, so to speak,
the acoustical lens through which they focus the high-frequency
sounds they emit to draw in the details of their world.

Research with beluga whales (*Delphinapterus leucas;* Red List:
Near Threatened 2008) reveals that the whales produce two
streams of sonar clicks separated in time by less than a second.
The twin beams are generated by phonic lips, intriguing lar-
ynxlike structures inside their heads that constrict air passing
through interior sinuses to vibrate surrounding tissues. Odontocetes control these vibrations with Pavarotti-like precision as

the sound passes through the head and into the melon. The less-than-one-second delay between one echolocational beam and its partner cancels the frequency of sound in some regions of space while enhancing it in others, effectively deflecting the beams to produce one wider ray of "light."

To the human ear—stripped of its own directional acuity in the underwater world—pilot-whale echolocation sounds like a buzzing beehive in three-dimensional overdrive, a nearly impenetrable wall of sound made all the more confounding by the fact that generally a number of whales are echolocating at once, producing layers of clicks as dense as white noise.

For the whales, however, this bath of sound shines a floodlight on their world—though the reflected images are not received in their eyes but in the parabolic bones of their skulls and particularly in the fat deposits in the long bones of their lower jaws. The images of their echoed sounds are also received in their ears, afloat in airspaces and fatty areas in their heads—so unlike our ears, tethered by bone to our skulls and unable to discern direction or most frequencies in the water. Odontocete echolocation not only illuminates the sea, it penetrates the sea and all things in it, right through to the bones.

The largest melon is found on the largest odontocete, the sperm whale (*Physeter macrocephalus;* Red List: Vulnerable 2008), and it contains something unique to this species, the spermaceti organ, full of a waxy liquid once thought to be congealed semen, hence the name. An old bull sperm whale's spermaceti organ might run one third the length of his body and contain three tons of oil.

This oil was the treasure lusted after by whalemen of old, an elixir for nineteenth-century medicine and technology, as ubiquitous as petroleum products are today and used for many of the same purposes: lubricants and rustproofing in watches, lenses, and scientific instruments; oils for transmissions and engines; waxy elements in lotions, ointments, glycerines, pomades, and detergents; additives to vitamin pills and medicinal compounds. Spermaceti was, in the words of the eminent whale researchers Hal Whitehead and Linda Weilgart, of Dalhousie University, Nova Scotia, the oil that lubricated the Industrial Revolution.[1] In its day it was also the finest fuel for candles and such a standard that candlepower—once an official unit of illumination—was calibrated against the light emitted in the course of an hour by burning a spermaceti candle.

Once upon a time, this waxy oil was also a culinary delicacy, a penchant guiltily described by Herman Melville in *Moby Dick:*

> The fact is, that among his hunters at least, the whale would by all hands be considered a noble dish, were there not so much of him; but when you come to sit down before a meat-pie nearly one hundred feet long, it takes away your appetite . . . But the spermaceti itself, how bland and creamy that is; like the transparent, half-jellied, white meat of a cocoanut in the third month of its growth, yet far too rich to supply a substitute for butter. Nevertheless . . . in the long try watches of the night it is a common thing for the seamen to dip their ship-biscuit into the huge oil-pots and let them fry there awhile. Many a good supper have I thus made.

In many odontocetes, males possess considerably larger melons than females, suggesting that they play a role in male-to-male aggression or female attraction or both. One theory posits that the enormous spermaceti organ of bull sperms acts as a bullhorn projecting the slow clicks, called clangs, unique to their species and gender. Whitehead and Weilgart propose that these clangs might be a form of sonic dueling by which males—who for most of their lives are out of sight of others of their kind—advertise their size, maturity, and physical and competitive fitness, a long-distance clanging of soundsabers.

Female sperm whales, in contrast, sing duets, most often in

close proximity, within sight of each other. Females inhabit long-term, even lifelong, associations, forming matrilines of related individuals: mothers with daughters, sisters, aunts, and all their offspring. Within these groups, females nurse and care for calves other than their own, and all will come to the defense of any calf and any other female within the family group . . . a fact the whalemen of old brutally exploited by harpooning, but not killing, a calf and then slaughtering the females who came to its defense.

Underwater recordings by Whitehead and his colleagues Tyler Schulz, Shane Gero, and Luke Rendell indicate that females carefully coordinate their songs to match those of their singing partners, copying the last phrase, or coda—composed of a short patterned click series—until both whales converge on a mutually agreed-upon repertoire, or riff. This jazzlike collaboration probably bonds whales already connected to each other, much as human song, once a staple of working and nonworking life, bonded people in families, tribes, gathering parties, hunting groups, churches.

In 1987, three years after Newfoundland, Hardy and I move on to film Hal Whitehead's pioneering work with sperm whales in the Galápagos, although we do not film Whitehead himself, as originally intended, since he is pacing impatiently in port awaiting a replacement part for his boat's engine. So we sail without him aboard a chartered yacht, conscripting one of his colleagues and a cobbled-together directional hydrophone like those Whitehouse and Weilgart use to track whales during their one-hour-long feeding dives thousands of feet deep. Hardy and I have arrived at the point in our filmmaking careers where we have a film contract, a budget, and a crew of six, with all the attendant premises, promises, worries, responsibilities, and deadlines.

It takes us more than a week to find our quarry in the deep

waters far offshore of the high volcanic islands of the Galápagos, sailing every day back and forth across the Equator until we bore Neptune himself. We mistakenly track many pods of pilot whales. We find orcas and huge schools of common dolphins. We are hindered by enormous seas and by our tiny, slow yacht, built in Newfoundland and as tight and airless as the inside of a whiskey barrel. It is an unseasonably hot El Niño year in the Galápagos, and we swelter in utter misery, unable to sleep whether we stay below and drown in our sweat or drag a sheet topside, only to have a passing rain shower chase us below again, a pattern repeated every thirty minutes.

Our skipper is a cantankerous old German with skin thickened by the scars of poorly removed tattoos that we imagine to be evidence of a Nazi past. One cameraman asks him about the tattoos during a sun cooker of an afternoon in the unsheltered cockpit. The old man's face erupts in a mass of uncontrollable spasms and twitches, and that is all he has to say about the matter. But soon thereafter he turns his whiskey-barrel boat back to port, a three days' sail. He quits us on the dock and anoints in his stead his son—an already accomplished young drunk with a crew of like-minded drunkards.

One sleepless night I notice a change in the motion of the boat and, peering out a porthole, realize we have drifted far from our anchorage beside one island and are about to run aground on another. Our young skipper has passed out on the bow, with his crew crashed around him like downed bowling pins. Another night, in the course of my insomniac wanderings, I swing my feet out of my bunk, determined to get some air topside, only to step down into four inches of diesel fuel sloshing belowdecks. Try as I might, I cannot rouse our inebriated skipper, the only soul aboard who can actually sleep.

We finally find the whales one afternoon—only moments after I've lost my temper with our demoralized conscripted hydrophone technician and accused him of absolute incompetence with this supposedly marvelous tool that is surely leading us *once again* to another school of pilot whales. He blinks hard and suffers my abuse . . . and then—there they are, right where he's led us, a pod of sixty or so female sperm whales and their calves logging, that is, lying motionless, at the surface during their midday nap.

It's like coming upon a herd of elephants in a bewitched sleep on the forest floor: the whales stretched out full length at the surface, their blowholes, canted to the left of their melons as is characteristic of sperm whales, dipping below the surface and bobbing up every minute or so. These sleeping breaths are unlike any other breath on Earth, the *anima divina* of colossi, an enormous draining of air through a small valve that opens with a cannon-fire pop and then empties the vast chambers with a wind that grows into a gale and then a tempest, whistling across the open blowholes before being sucked back in again, a tornado run in reverse. The volume of air visibly expands the whales' bodies and affects their buoyancy: they rise higher above the waves during their inhales and drop nearly from sight during their exhales. Twenty whales buck the seas and suckle the sky in private storms of their own making.

Now the directional hydrophone proves its miraculous abilities, enabling us to follow the whales when they awake and begin their feeding dives. We cannot see them for the thirty to sixty minutes they spend in the utter darkness of the mesopelagic (Greek *mesos:* middle; *pelagos:* open sea), between 1,000 and 3,000 feet deep. But through the secret imagery of their sounds, we can imagine the whales' bodies being squeezed by the pressure as they hunt bioluminescent squid with the lure of white jaws bathed in the bioluminescent mucus left over from prior squid meals.

Thanks to the directional hydrophone we can sometimes

determine where the whales will rise from their dives and can therefore witness the tremendous spectacle of a whale ascending at such speed as to burst headfirst through the surface like a Poseidon missile, launching nearly clear of the water before falling back to sea. Most of the time we know where the whales are coming up because they have left a small calf or two at the surface, tended by one adult or adolescent after another, the whales staggering their dives to accommodate babysitting duties.

But even with the directional hydrophone, we still lose the whales as often as we maintain contact with them. Sometimes they simply go silent and slip away. At night they dive harder and more frequently, and we can't easily keep up. Once we hear orcas on the hydrophone, and the sperm whales fall instantaneously, eerily silent and, as far as we can tell, simply disappear, as is clearly their intent.

Our adventures in the Galápagos take place a decade before Robert Pitman and Susan Chivers observe thirty-five orcas, or killer whales (*Orcinus orca;* Red List: Data Deficient 2008) attacking nine female sperm whales off California.[2] Nothing of its like had been witnessed in modern times, and modern whalemen, that is, researchers, had supposed that sperm whales were invincible in the sea.

But all that changed when Pitman and Chivers came upon an assault already in progress. They observed for more than three hours as wave after wave of female killer whales attacked the sperms, inflicting great injury before withdrawing, leaving the larger whales to bleed and weaken before attacking again—and again and again. Throughout the battle, the sperms put up no defense other than to form their characteristic protective rosette formation: gathered at the surface, heads pointed in to the center, tails fanning out like the spokes of a wheel. Pitman and Chivers described the scene in *Natural History:*

During one of their sorties, a sperm whale is pulled away from the rosette and immediately set upon by four or five [female killer whale] attackers . . . Twisting their bodies and violently shaking their heads like huge hungry sharks, the killer whales try to wrench off mouthfuls of what must be very tough flesh . . . Then, to our astonishment, two sperm whales leave the rosette formation and approach their isolated companion. One on each side, the two begin to herd the severely injured whale back to the rosette. For a time, the killer whales redirect their attack to the escorts, then retreat once again. We see this same heroic scenario several times . . . Eventually the sperm whales become disoriented. They try, but fail to hold the rosette formation. All appear to be wounded, several severely . . . Several sperm whales have been dragged away from the rosette and are being savagely attacked. One of the largest rolls slowly over on its side like a sinking ship and appears to be very near death. Then, as if on cue, a bull killer whale rushes in. He broadsides the isolated sperm whale, pushing it sideways through the water. Like an angry dog, he seizes it by the flanks and shakes it violently from side to side, then swings it around in an arc, throwing up huge sprays of water . . . The actions of the female killers have been demure compared with the power exhibited by this animal.[3]

Pitman and Chivers believed that all the sperm whales sustained lethal injuries that day. One was disemboweled, another partially skinned alive. The orcas killed far more whales than they could possibly eat, a fact that greatly puzzled the observers. Sailing away, they watched as the dying whales struggled, with little success, to regroup into their rosette formation.

25

Salting Down the Lean Missionary

TWO WAVES OF HUMAN whaling in the past three hundred years reorganized the universe of sperm whales. The first wave was a purely American invention, born about 1712 in the waters off New England. Its bounty at least partially sustained the thirteen colonies during their War of Independence against Great Britain. The sperm whales who fell victim to this open-boat hunt fueled and lighted the world during an era that might as well be known as the Spermaceti Age, which lasted one and a half centuries and killed nearly 30 percent of an estimated global population of 1,100,000 sperm whales.

For sixty years, beginning in about 1880, the whales were offered something of a respite—only to have it viciously repealed in the aftermath of World War II, when all the advantages of the Petroleum Age were brought to bear upon the survivors of the Spermaceti Age. Speedy ships and exploding harpoons enabled a level of hunting never seen before, quickly reducing the global population of sperm whales by an estimated 67 percent to about

360,000 individuals. The carnage of the second wave of sperm whaling lasted only four decades yet killed three times as many whales as had been killed in the entire sixteen decades of the first wave. It came to an end only in 1985, when the International Whaling Commission granted a moratorium—a bitterly hard-fought ceasefire that has continued to be attacked during every subsequent annual meeting of the commission to this day.

The ecological aftereffects of these waves of hunting may explain something of the terrible success enjoyed by the thirty-five killer whales attacking the nine sperm whales in 1997 off the coast of California, only twelve years after the moratorium. The second wave of human whaling not only killed far more sperm whales far more quickly, it also far more effectively targeted the bull whales summering in the highest latitudes on the edges of the polar ice both north and south.

In these remote waters, the males sequester themselves to grow to their legendarily huge sizes—up to 60 feet and 60 tons—as compared to the females' 33 feet and 13 tons. Sexual dimorphism (Greek *di:* two; *morphe:* form), common among cetaceans (Greek *kētos:* whale), including dolphins, reaches its zenith among sperm whales.

The bulls earn their girth during the first twenty-five or more years of their lives, and this task apparently requires their near monastic seclusion from the females and the presumably cold-sensitive calves. Or the genders may occupy different waters so as not to compete for food sources. But whatever the reason, after leaving their natal groups, males join bachelor schools and move to colder waters. As they age and grow, they move progressively farther apart, their bachelor schools becoming smaller as the individuals grow bigger, until the largest bulls live entirely alone in the highest latitudes, some in the waters around Newfoundland. These solitary giants venture toward the equatorial realm of the females only in their late twenties, finally to affiliate themselves with one or more groups of females for mating.

And so it's possible that the attack witnessed by Pitman and Chivers was absent one or more important players who might have cast an entirely different plot line upon the drama. Whitehead and Weilgart and others observed a killer-whale attack near the Galápagos in the 1980s in which a group of female sperms clustered tightly together while maneuvering to keep their heads and jaws facing their attackers—exactly the opposite of what Pitman and Chivers witnessed. The Galápagos whales were also protecting a small calf in their center. Meanwhile a large and truly invincible bull sperm whale was in attendance with the females and took up a position behind them, safeguarding their rear. This layout enabled the sperms to successfully defend themselves during a nearly three-hour-long killer-whale attack.

To this day, in the aftermath of human whaling, large male sperms remain uncommon on many breeding grounds—particularly in the South Pacific, where adult males were severely decimated—and their scarcity may affect more than just the ability of females to protect themselves against killer whales. Whitehead and Weilgart suggest that normal birthrates, which are now unnaturally low in sperm whales, may be dependent on the presence of more than one large bull among a group of females. Perhaps the clanging of soundsabers is needed to stimulate the males to breeding activity or the females to breeding activity or both.

Something like this occurs in humans, who share with sperm whales a relatively low rate of conception per copulation, and for whom frequent copulation is as much sociable as it is reproductive. Human men do not duel long-distance with soundsabers, or at least did not before the Internet Age. But studies of human male responses to pornography indicate that they are more aroused sexually and produce more sperm when they experience even fictional threats of competitors—in this case, after viewing films of sexual activity between one female and multiple males.

Herman Melville's character Ishmael disparaged the habit among old Nantucket whalers of eating whales: "That mortal man should feed upon the creature that feeds his lamp, and, like Stubb, eat him by his own light, as you may say; this seems so outlandish a thing . . ."

> But no doubt the first man that ever murdered an ox was regarded as a murderer . . . Go to the meat-market of a Saturday night and see the crowds of live bipeds staring up at the long rows of dead quadrupeds. Does not that sight take a tooth out of the cannibal's jaw? Cannibals? who is not a cannibal? I tell you it will be more tolerable for the Fejee [Fijian] that salted down a lean missionary in his cellar against a coming famine . . . than for thee, civilized and enlightened gourmand, who nailest geese to the ground and feastest on their bloated livers in thy paté-de-foie-gras.

Despite the worldwide moratorium on whaling, Japan continues to kill whales, including ten sperm whales a year under a special permit euphemistically described as "scientific whaling"—whose real purpose is to keep Japan inside the regulatory body and not roaming piratically outside. The truth is that this "cultural legacy" of hunting whales generates less than one one-thousandth of the gross domestic product of the world's second largest economy—a drop in the oil-pot that not even Stubbs would deign worthy of consumption. At any rate, few Japanese eat whale meat anymore, which carries enough mercury and organochlorine pollutants to make human consumption of it worrisome. Much scientifically whaled whale meat is now used in Japanese dog food.

Japan defends its whaling industry with claims that sperm whales compete with human fishers. It is true that all the world's sperm whales combined consume a biomass similar to all the human marine fisheries of the world combined, roughly 100 million tons a year. Yet the diet of sperm whales consists almost entirely of mesopelagic squid weighing about two pounds, a market barely exploited by human fishers. Clearly, the removal of sperm whales—or any whales—from the sea runs the same risk of triggering trophic cascades as removing cod from Newfoundland waters.

Ishmael bracketed the breadth and narrows of human thought on these matters, first when he questioned "whether owing to the almost omniscient look-outs at the mast-heads of the whaleships, now penetrating even through Behring's straits, and into the remotest secret drawers and lockers of the world . . . Leviathan can long endure so wide a chase, and so remorseless a havoc." He concluded:

We account the whale immortal in his species, however perishable in his individuality. He swam the seas before the continents broke water; he once swam over the site of the Tuileries, and Windsor Castle, and the Kremlin. In Noah's flood, he despised Noah's Ark; and if ever the world is to be again flooded . . . then the eternal whale will still survive, and rearing upon the topmost crest of the equatorial flood, spout his frothed defiance to the skies.

Nevertheless, Ishmael knew that even defiant leviathans suffer extinction. Melville published *Moby Dick* fifty-five years after Georges Cuvier presented his seminal paper to the National Institute in Paris describing the fossil mammoth bones that finally proved beyond scientific doubt the "existence of a world previous to ours, destroyed by some kind of catastrophe." Cuvier deemed this notion of extinction, this changeover of species, a revolution. Its proof came sixty-four years before Charles Darwin's *On the Origin of Species* outlined the mechanism by which species might arise.

Yet even without knowledge of evolution, Melville and his character Ishmael certainly knew that everything falls away.

Aboard our whiskey barrel of a boat during a steaming El Niño, we begin to dream of joining the sperm whales in the cool plankton-green of the equatorial Pacific. Few humans other than Whitehead, Weilgart, and their crew have ever willingly done this. But we are tempted by the mystery of the whales and by the lure of filming what has not been filmed before.

So we slip overboard with snorkels and fins and waft as gently as jellyfish toward a distant group of resting females and calves. Moby Dick is foremost in our thoughts. Sperm whales, renowned for their courage and their temper, have been the cause of many of the whale-wrecked boats of the past and present—almost all, certainly, a result of the whales fiercely defending themselves or their kin. But we cannot be sure what constitutes a threat to a sperm whale and whether our measly squid-sized bodies, which can hardly swim and which have been disgorged from a vessel the likes of which some of the females surely remember once hunted their kind, pose enough of a threat to arouse their wrath.

So we proceed slowly and gently, wishing for the transparency of jellyfish. Nevertheless, we have to swim hard against the current and what we eventually realize is an additional current set up by the whales lazily fanning their enormous flukes and pushing a backwash toward us. With the infinite languor of the unconcerned, they are swimming away from us. We paddle harder, sweating underwater and grunting through our snorkels, knowing even as we do so that it's a waste of time and energy, that we cannot hope, even from within the bubble of this dream of swimming with sperm whales, to keep up with or catch them. As if to drive that point home, the whales begin to editorialize, as cetaceans are wont to do when humans follow in their wakes, and we swim into enormous blooms of brown whale shit.

Still, we keep at it. We redesign our strategy, climb back aboard the boat, motor a huge circle far around them, and get dropped off ahead of the whales, avoiding their unsalable backsides. Again we waft in, only to find ourselves having to swim hard and then harder as we realize that the whales have ever-so-slightly flexed their flukes and shifted course by a degree or two—just enough that once again, even from within this perspiring bubble of a dream, we cannot catch them. Once, per chance, on my swim back to the yacht, I look down and see a sperm whale twirling playfully under the keel.

We follow the whales for a few days without success until we finally add the missing ingredient that has worked for many another cetacean species: sound. Although the sea is full of the clicking codas of the female sperm whales talking to each other, Hardy begins to whoop and gurgle underwater through his snorkel in feeble semblance of a humpback whale. We all join in. No longer do we have to swim hard through whale latrines. Within seconds, the pod has switched direction to face us, stalled in the water on the brakes of its curiosity.

It's surely the most extraordinary sight any of us has ever witnessed. Suddenly, in the deepest of oceans far offshore of the Galápagos Islands, the whales we have been dreaming of are driving straight for us, full steam ahead. Their huge rectangular melons, with underslung white-rimmed jaws, resemble nothing so much as whitewalled boxcars with flukes for engines, and unexpectedly we are on the tracks and their freight train is headed our way. Somehow I have ended up at the front of this whooping human enterprise. Steaming at me with all the energy and incaution of adolescence is a young male sperm whale. He does not appear to have any intention of stopping, and I believe he aims to ram me. I begin to backpedal with my hands, the most pathetic of human responses, only to realize the futility and stop—consciously deciding to observe this moment of my death rather than waste it panicking.

The whale surges toward me, his boxcar so huge that I have to look up and down to take it all in. I can see the asymmetrical bulge of his blowhole. I cannot see his eye, set too far back on his head. I am awaiting my fate with resignation, some interest, and, surprisingly, no fear, when—only a few feet away from me—he suddenly bows his great head, an almost courtly introduction, and begins to slip under me and just off to my right side. I realize that I am about to learn the answer to my long-standing question: what would happen to a person in the water if a whale sounded directly alongside her—would she, like a person afloat beside a sinking ship, be dragged under too?

He jackknifes his huge head downward, and I can see the sheets of cellophane-thin gray skin peeling off his body—the constant striptease one of the means by which cetaceans reduce their drag in the water. I can see the sculpted angle of his cheek and jaw, tapered like a keel. I can see his eye, inside the heavy lid, rolling back and forth, up and down, examining me. And then his head has slipped below me, and the huge submarine of his body slides past, the wrinkled aft quarters, the perfectly tucked paddle of his pectoral fin, the rounded dorsal fin. All the musculature of his carved muscles bunches up alongside me, close enough to touch—and I do, feeling his unimagined softness and his warmth. Then the huge stock of his tail rises into the air. I lift my head, water streaming down the faceplate of my mask, to see the incomprehensible umbrella of his flukes, shading me from the equatorial sun, shivering with their own tensile strength, before arching upright and slicing into the depths. He descends with breathtaking speed to a pinpoint of darkness, and all that's left of him are the small spiraling eddies of water rising in his flawlessly hydrodynamic wake.

He has not salted down this lean missionary, and I am grateful.

26

The Existence of a World
Previous to Ours

T HE CURRENT GLOBAL population of great whales of all
the major species combined—right, bowhead, blue, fin, sei,
Bryde's, humpback, gray, minke, and sperm—is estimated at
somewhere shy of 500,000 individuals. To say there were once
many more seems grossly inadequate, and a recent study implies
just how staggering the loss of historical abundance is. Calibrat-
ing genetic variation—one marker of the original size of a population—Stephen Palumbi of Stanford University and Joe Roman
of Harvard University estimated the combined historical populations of only three species of whales (humpback, fin, minke)
inhabiting only the relatively tight quarters of the North Atlantic
at just shy of a million—twenty times higher than the present-
day populations.[1]

The authors do not extrapolate. But if we do, it becomes at
least theoretically conceivable that today's 500,000 whales were
yesterday's 10 million whales. Jeremy Jackson of the University of
California, San Diego, proposes that prior to large-scale human

hunting of the great whales, predatory fish, and sea turtles, the biomass pyramids of the sea were at least partially inverted. The profile we are accustomed to seeing—huge foundational tiers on the lower trophic levels supporting ever-smaller tiers of higher-trophic-level predators—was reversed in marine ecosystems. Jackson's work with green sea turtles (*Chelonia mydas;* Red List: Endangered 2004) suggests that some 91 million adults of that species inhabited the tropical western Atlantic in the late seventeenth century—comprising a larger biomass than all the bison in North America prior to their decimation.[2]

We cannot easily reconstruct the trophic cascades that surely followed the loss of so many large oceanic consumers, though intimations exist. James Estes, Daniel Doak, and Terrie Williams, of the University of California, Santa Cruz, propose that the deficiency of great whales in the Bering Sea has triggered a severe and ongoing decline in Steller sea lions (*Eumetopias jubatus;* Red List: Endangered 2008) and sea otters (*Enhydra lutris;* Red List: Endangered 2008)—who now fall prey to killer whales forced to hunt smaller animals in the absence of their favored prey, the great whales.[3]

The North Atlantic right whale (*Eubalaena glacialis;* Red List: Endangered 2008) was the first whale to be extensively hunted by humans, as early as or even before the eleventh century, by Basque whalers in the Bay of Biscay. Possibly predating Christopher Columbus's arrival in the New World, but certainly by 1530, the Basques were chasing right whales across the Atlantic to the shores of Newfoundland and Labrador, killing some tens of thousands there (though recent investigations suggest these might have been bowhead whales). Right whales were targeted and named because most floated after death, making them the *right*—most manageable—quarry for men rowing to the whale hunt. The species was also right because it was slow moving and relatively docile, a combination of traits that ensured its complete extinction in the northeast Atlantic. The northwestern Atlantic population is today estimated at a mere 300 to 350 individuals.

Darwin perceived the dangers in underestimating the threat of diminished populations when he wrote: "To admit that species generally become rare before they become extinct—to feel no surprise at the rarity of a species, and yet to marvel greatly when it ceases to exist, is much the same as to admit that sickness in the individual is the forerunner of death—to feel no surprise at sickness, but when the sick man dies, to wonder and to suspect that he died by some unknown deed of violence."

We can imagine something of the trophic revolutions attending the loss of so many great whales—the deepest and darkest parts of that loss. Countless millions of behemoths once churned the sunlit waters of the world, some scouring the sea floor for buried mollusks, some chasing capelin up from the deep, some corralling krill, some wrestling giant squid. Yet these cumulative actions of their daily lives in no way marked the sum of their services in the world, which persisted far beyond the span of their own existence into the fetch of the future. Like old-growth trees that fall and die and spend centuries nourishing a forest with the fruits of their rot, so the great whales sink to the bottom of the sea to nourish the oceans with their decay.

A great whale's death journey to the bottom of the sea is accelerated or braked by the presence or absence of fat in its body. Lean, starved whales fall faster than fatter whales. Regardless, the downward journey takes place in the slow motion of the underwater world, as the processes of decomposition produce buoyant gases that duel with the force of gravity in such a way that the carcass rides a gentle elevator up and down on its way down. A sperm whale off the Galápagos, a gray whale in the Gulf of California, a humpback off Newfoundland, all become part of the windfall that descends from the surface to the collapsed depths of the abyss.

The first visitors to arrive at the deathbed of most whales are the swimmers of the pelagic ocean, the torpedolike mako sharks, blue sharks, silky sharks, tuna, and billfish, who tear at the carcass, opening it for the big fish with small mouths and for the smaller fish with big mouths and for all the tiny planktonic predators who will join the dead whale on the elevator ride up and down. Later the bulkier scavengers, tiger and great white sharks, will work their way across scores or even hundreds of miles of ocean to join the whale's vertical migration.

In the ink-black waters below 600 feet, the mesopelagic community begins to attach itself to the carcass of a great whale—the barracudinas, the cross-toothed perches, the myriad of squid, the lanternfish and lancetfish. When a falling whale enters their world—a world the whale might well have fed upon her entire life—those who were her prey gather to feast upon her remains. Many of these dwellers of the middle sea possess extraordinarily large eyes and a series of light-producing organs known as photophores. Hidden within perpetual darkness, these fish produce weak blue, green, or yellow lights, whose colors and patterns reveal the individual's species and gender, as well as information shared for the purpose of shoaling and other communications we cannot understand. The photophores, controlled by internal dimmer switches, also create a camouflage known as counter-illumination, so that those dining on the falling whale are as good as invisible against the faint glow from above.

Pushed and prodded by a ravening army of razorblades and cutlasses, the whale slips into the bathypelagic (Greek *bathus:* profound, deep) zone below 3,000 feet. Here the pressures of the depths are greater than one hundred times those at the surface, squeezing the buoyant decomposition gases from the tissues of the whale once and for all, so that now her inexorable path is downward only. The carcass begins to fold and even break at the weak joints, the first stages of the flattening of the deep.

New fish appear, species created in Ægir's most phantasma-gorical moments: animals missing parts seemingly vital to existence in our world, such as hearts, brains, kidneys, muscles, scales, eyes, gills, skeletons. Some of these dwellers of the profound sea are endowed with the ability to eat three times their body weight, stuffing whale meat into distensible stomachs until their bellies dwarf their bodies, as if upside-down hot-air balloons were dangling below the tiny gondolas of their backs and fins. Some, like gulper eels, become little more than floating jaws with cellophane bags for bodies—a plan designed to maximize the one and only meal that may come their way in a year.

Odd gelatinous anglerfish approach, the females casting with the fishing tackle attached to their foreheads, complete with rods, lines, luminescent lures, and sometimes hooks. Their male partners take the form of tiny parasitical appendages attached by their jaws to the females' skin, where they subsist on embryo-like connections to her blood supply. Thus conjoined, the males atrophy to nothing more than gonads, awaiting the moment to spawn.

Everything arises, everything falls away.

Finally, hours or even days after death, the whale finds the soft bottom of the sea and settles into the muck, where it becomes an island for the migrations of sleeper sharks, king crabs, and hagfish. These scavengers are joined over time by the slower-moving migrant feeders of the abyss: sea cucumbers, sea stars, brittle stars, burrowing urchins, and hemichordate worms, who caravan between the feeding stations of large things fallen from the surface—the big whales, big fish, big tree trunks, and huge mats of kelp and other vegetation blown offshore by storms.

On the vast desert of the ocean floor, a whale fall is an unexpected oasis literally dropping out of the blue and providing a nutritional bonanza of a magnitude that might otherwise take thousands of years to accumulate from the background flow of small detritus from the surface. A 35-ton gray whale takes one and a half years to be stripped to the bone by the scalpels and stomachs of the deep. A 160-ton blue whale might take as many as eleven years. Yet even beyond that, even after a dead whale's organic matter appears to have been completely devoured, the blubber-saturated sediments surrounding the bones support dense populations of chemosynthesizing bacteria and archaea. These active sediments support vibrant communities of nematodes, foram protists, bivalves, and worms.

When nothing is left but the skeleton, the bones themselves become fuel for a preternatural whale-fall community, including *Osedax,* or bone-eating, worms, which consist mostly of roots that bore into the skeleton to extract its proteins. Bone-eating worms are females—the dwarf males form harems inside the tubes of larger females—and are aided in their work by symbiotic single-celled bacteria able to fuel themselves on the organic carbon of the dead whale. Thanks to these bacteria, bone-eating worms survive without any gut whatsoever. And they are only one example among the queer wonders of abyssal oases, where much of the primary production is done by chemoautotrophic (Greek *chemo:* chemistry; *autos:* alone; *trophe:* food) organisms that make life from nonlife, using chemistry not sunlight.

In a world without eyes, unseen by any living thing, the dead whale becomes a shadow atop the shadow of the bottom of the world, living in its afterdeath for another hundred years, maybe more.

The whale-fall researchers Craig Smith and Amy Baco estimate that even with today's remnant populations, 69,000 great whales die every year, most sinking into the depths to join an abyssal community of perhaps 850,000 "living" whale falls.[4] That seems a great richness of whale falls. But imagine the existence of a world previous to ours, when perhaps 10 million whales roamed the deep blue home and had done so for countless millennia. Surely hundreds of whales might have died every day in the World Ocean in the time before human whaling. If so, perhaps millions of whale falls once enriched the abyssal plains, transforming the desertscape of the deep into fruitful orchards of the dead.

Whale falls are known to host unique communities of animals, with some overlap of species between similarly organic wood falls and kelp falls. It's conceivable that whale falls also help seed other extremophile communities inhabiting the deep sea: the hydrothermal vents and the cold seeps scattered throughout the abyss, those other unordinary places where life adapts to extreme pressure and a sunless world.

Sea-floor hydrothermal vents, some at depths of 10,000 feet, are the equivalent of underwater pressure cookers squeezed by forces three hundred times stronger than those you and I feel at sea level. They form at fissures torn into Earth's crust by our planet's persistent rebuilding of its house, by its slow-motion rearrangement of ocean basins and continents. Much of this building takes place down the spine of the longest mountain range on Earth, the 40,000-mile underwater chain threading through the center of all the ocean basins and connecting one to the next.

Hydrothermal vents form in submerged alpine zones where the sea floor is spreading to form rift zones. At some places ocean water is drawn into these vents to mix with magmatic water released from the upwelling of molten rock. The mix is hot, sometimes very hot, upward of 760 degrees Fahrenheit, a veritable cauldron amid a near freezing background of 35-degree seawater. Accordingly, hydrothermal vents are closed systems—aquariums, really, contained by walls built by temperature differences and by minerals emitted from the vents: sulfides, barium, silicon, calcium. They support highly specialized and uniquely adapted species that exist in relatively low diversity but in enormous plenty, up to 100,000 times the density of life forms on the surrounding deep-sea floor—bioabundance in an invisible box.

The first hydrothermal vent community was discovered only in 1977, and the first whale fall only a decade later in 1987. And yet from the beginning, researchers have wondered if there might be a connection between these two types of abyssal oases. Could the relatively abundant and long-lived whale falls act to seed other communities? Could the larvae dispersed from the whale's community of invertebrates or vertebrates find their way along the underwater rivers of the world to other isolated extremophile environments? Genetic studies of mussels from the subfamily Bathymodiolinae suggest that some who colonize whale falls, kelp falls, and wood falls also colonize vents and seeps, perhaps helping shape the community of life at these biological outposts in the course of the long clock of evolutionary time.

Beyond that, the nutritional bonanza from whale falls plays a reverse role as well, with at least some of the dissolved organic matter of the dead whale and its dead community riding the vertical conveyer belt driven by the winds blowing across the surface of the sea to create upwellings. This rich broth rising from the abyss and fertilizing the phytoplankton in the sunlit zone makes the surface an even more bountiful environment. Surely the presence of millions of whale falls in times past enriched the carbonated surface beyond our imagining, seeding it with the strength of giants or of *jötnar*—millions of bottomless tankards of Ægir's beer.

27

Reading God

ONE OF SCIENCE'S PRESCIENT ironies is that a good many of today's good-news nature stories are a result of researchers scrambling to catalogue Earth's life forms before they disappear. Breaking through the daily clutter of war news and economic news and political news are the quiet announcements of new species, including truly startling finds: a new catlike mammal in Borneo, a new songbird in India, a snake that changes colors, and, perhaps most astounding, an entirely new phylum—one of the highest taxonomic divisions of animals—of wormlike creatures. Most extraordinary are the marine discoveries, including an average of three new species of fish a week since 2000.

The discovery of novel oceanic life forms comes thanks to the Census of Marine Life, a network of two thousand researchers in more than eighty nations engaged in a ten-year initiative to assess ocean life and its changes over time. The Census is a veritable factory of revelations: species surviving below 2,300 feet of ice off Antarctica; a living shrimp that had been believed extinct for 50 million years; pairs of mated seabirds flying 44,000-mile figure-eight loops in the two hundred days between nesting seasons; a fish school 20 million strong, covering an area the size of Manhattan, swimming off New Jersey.

The quest reveals a startling truth about our planet: that life thrives even in "lifeless" conditions, harvesting energy from sources we thought were ours alone—radioactive uranium, hydrogen, hydrogen sulfide, and methane—in places no one thought to look before, including between rock crystals. Some archaea, bacteria, and fungi live more than two miles below Earth's surface. Some cyanobacteria in Antarctica escape the weather by living inside quartz that is clear enough to allow photosynthesis. We call these organisms extremophiles because their worlds are uninhabitable by us, although many dwell in conditions that mimic what we know of the first life on Earth—in the eons before photosynthesizing plants freed the poison of oxygen from its bond with water to grow an atmosphere that would eventually nourish the truly extremophile aerobic life forms that eventually led to us.

On Hydrate Ridge, fifty miles off the coast of Oregon, the mud on the bottom of the sea is more alive than dead. Thousands of feet below the surface, under pressures that would crush you and me, in temperatures that would turn our blood to sludge, worms, crustaceans, snails, bacterial filaments, and other life forms as yet undescribed thrive in an environment largely devoid of oxygen and toxic with hydrogen sulfide and methane. This is the world of a cold seep, a sea-floor ecosystem discovered only in 1984, where hydrocarbon-rich fluids bubble from the bottom at cold, nearly ambient water temperatures, varying with depth and latitude and supporting unique biological communities fundamentally different from our own.

Seep life is driven not by photosynthesis but chemosynthesis, whereby microorganisms use methane and hydrogen sulfide to fuel a gas-powered food web. The byproducts of this process create underwater reefs of carbonate rocks, something like coral reefs, that shelter otherworldly communities. Seep tubeworms from the Gulf of Mexico can live 170 years or more. Scattered erratically across the bottom, these cold seeps and their cousins, the hydrothermal vents, mark the ocean's farthest-flung and most alien outposts.

It's 2006, nineteen years after the Galápagos and twenty-two years after Newfoundland. I have left the world of documentary filmmaking behind for the carefree life of a sea tramp, hitchhiking on whatever scientific cruise will take me, my backpack, notebook, foul-weather gear, and steel-toed deck boots. Once again I can, if I wish, travel without budgets or contracts, without the twenty-plus cases of film gear, the film crew, and most attendant premises, promises, worries, responsibilities, and deadlines. I've come nearly full circle from my days on Isla Rasa twenty-six years earlier—in migration much of the time with only a few dollars in my pocket and only a vague notion of how I'll get home again.

At present I'm tagging along on a trip to Hydrate Ridge, where science requires big tools—in this case, the hard-working, far-sailing research vessel *Atlantis*—to study tiny beings, the minute creatures inhabiting the mud of the sea floor. The mud stinks of rotten eggs and flatulence and comes aboard via the submersible *Alvin,* which cruises 2,500 feet below in a world so remote and so unwelcoming that only three people at a time can get to it in undersea assignments as coveted as space-shuttle berths. In this undersea mud live extremophile life forms invisible to the naked eye—though mind-bendingly huge when viewed through the extended eye of the dissecting microscope—including translucent worms whose blood cells can be seen chasing each other through their vessels and miniature crustaceans that look like the Alien from *Alien.*

Weather permitting, the submersible dives each morning carrying two researchers who spend the day in near fetal position staring out portholes the size of grapefruit and breathing recirculated air infused with the smell of their own dirty socks. While the scientists work their task lists, the pilot manipulates *Alvin*'s robotic arms, biopsying holes in the bottom and collecting mud from around the seeps. Jules Verne could hardly have imagined the wonders teased, threatened, or delivered on every dive. "Two crystal plates separated us from the sea . . . Who could paint the effects of the light through those transparent sheets of water, and the softness of the successive gradations from the lower to the superior strata of the ocean?"

Roughly seven hours after launch every day, all three aquanauts return to the surface looking blissful from their time in the deep but also aged a hard decade by fatigue and the inescapable dehydration aboard their sealed miniature *Nautilus*. With only a piss bottle for a toilet, most submariners are reluctant to drink much if any liquid on dive day.

The first crew to dive at Hydrate Ridge on this voyage includes Lisa Levin of the Scripps Institution of Oceanography, a renowned expert on the strange chemical worlds of cold seeps and the chief scientist for one of the two research teams sharing the cruise. Her crew consists of twelve marine biologists and biogeochemists collaborating on a study of seep ecology and evolution. Petite, wearing jeans and Converse sneakers, Lisa is perpetually on the move, leaning forward in pursuit of questions, leaning back to issue commands, fast and quiet, mud-spattered, clipboard in hand. She doesn't say much, maybe the result of working the deepwater for twenty-seven years, of making seventy dives into a quiet world no humans have words for.

She loves the bottom. It's possible to read mud, she says. I picture her peering out through *Alvin*'s porthole and tracking the infinitesimal movements of clams and sea stars.

Lisa's group is studying bristleworms, and in particular, why so many bristleworms from the family Dorvilleidae inhabit these seeps—seventeen species so far, including ten in one genus. Dorvilleid worms appear to be the underwater equivalent of the Galápagos finches, which proved so seminal in Charles Darwin's thinking on evolution. From them he stitched together the startling revelation that all thirteen finch species inhabiting the Galápagos had radiated from a handful of ancestors blown there from South America and had eventually evolved into a constellation of new species resembling blackbirds, wrens, grosbeaks, woodpeckers, and mockingbirds more than finches.

The dorvilleid worms may be reacting similarly, having arrived at the abyssal islands of underwater seeps and whale falls to build ephemeral communities lasting the decades or centuries that fuel is available. Like Darwin's finches, the worms are rapid responders, filling empty ecosystems with fully functioning communities in no time at all, evolutionarily speaking. They offer examples of the kind of speciation that likely followed on the heels of prior mass extinctions, as survivors made creative and speedy use of divergent evolution to adapt to newly vacant niches.

In Lisa's lab aboard *Atlantis* I'm learning to sort worms. Staring at tiny living creatures magnified twelve or twenty-five times in petri dishes full of deoxygenated water has been a hypnotic through-the-rabbit-hole exercise: foraminiferans like frosted Christmas-tree ornaments; spaghetti worms like exploded dissections gone bad; snails wearing circa-1960 bathing caps with bobbing bacterial filaments, all in a snow-globe blizzard of gold, mica, and dragonfly-iridescent flakes. We work with paper-thin forceps and glass pipettes. Yet many of these creatures are so delicate—notably the tiny translucent bivalves—that they simply disintegrate when touched.

The seas have kicked up today, and some of the researchers normally glued to stools along the microscope bar have retired to ride out the bad weather in their bunks. Through my looking glass, the mud world in the petri dish reels from the mini-tsunamis triggered by the rolling of the ship — worms flailing, snails spinning in shells, amphipods sprinting on swimmerets, bacterial filaments tying themselves in knots. It's a stomach-churning chaos presumably unknown in their deepwater home.

The truth is, I'm beginning to have qualms. I feel sorry for the worms, some of whom, I imagine, may be more than 170 years old. My doubts crystallize when Ken Halanych, a marine biologist from Auburn University, tells me I have to dissect the worms that are still in the tubes they've built, not just collect the ones floating free. This is all part of the sorting business, of course, seeing who lives where, correlating it with gas measurements of the mud layers they inhabit, quantifying the ecosystem and searching for new species. Halanych is my friend at the dissecting scope, one of the only scientists who seems enthusiastic about the hours I spend there and willing to teach me. But I can't do what he wants.

Even though all these worms brought up from the bottom are destined to die anyway, to sacrifice themselves for science, as the graduate students joke, I can't bring myself to yank them from their glovelike tubes. So I leave the dissecting scope and return to my more mindless assigned tasks of labeling Nalgene sample bottles and assembling tube cores for biopsying the sea floor, for collecting the plugs of mud and all the life within them.

Later I overhear Victoria Orphan, a serene geobiologist from the California Institute of Technology and the chief scientist of the other team aboard *Atlantis,* confess her own squeamishness about worm dissections. Maybe because of this, her chosen work is with the methane-eating bacteria and archaea that fuel the gas-powered food web at the seep. She observes these invisible (to us) creatures through chemistry alone, death throes mercifully invisible.

Victoria's study addresses a pressing issue in the biodiversity crisis. In past warming eras, vast undersea deposits of methane ice may have melted, burping gas through the water column and into the atmosphere to cataclysmically accelerate global warming. Some scientists theorize that a methane burp precipitated — or at least contributed to — the Permian-Triassic extinction event 251 million years ago, in which nearly all life on Earth perished.

The fact that little of today's methane seeping from the sea floor ends up as a greenhouse gas in the atmosphere may be thanks in part to the microorganisms inhabiting cold seeps. Victoria's bacteria and archaea may help keep our world livable — a reminder of the power of the meek, including the unknown power of the undiscovered meek.

For a while I switch to Victoria's lab. It's younger, noisier, more crowded, less reverent. The graduate and undergraduate students on her team name their tube cores for rappers and discuss the quantity of bling in their mud. These young scientists are powerfully, almost alarmingly, focused on their work, barely breaking from it to take note of schools of dolphins on our bow, humpback whales, soaring albatrosses — which they refer to, wistfully or dismissively, as charismatic megafauna.

I wonder if anywhere along their educational route they are encouraged to consider science's own footprint. The cold seeps we are visiting are frequented by teams from all over the world. Every submersible drops its ballast before ascending to the surface, and popular dive sites are paved with metal brick weights. When I mention this, various researchers assure me the metal will rust away in seawater. I point out that the seeps are anoxic environments, nearly devoid of rust-inducing oxygen . . . Like Mount Everest, Hydrate Ridge is littered with the all-too-durable refuse of exploration.

A week later, at Eel River Basin seeps off California, the *Alvin* brings up big slabs of rock punctured with deep, perfectly round holes. The two returned scientists tell me of scores of holes as far they could see down there. Everyone is excited by the doughnut rocks. There's speculation about bubbling methane "chimneys." Yet somehow the holes look familiar to me. They remind me of the tube cores I spend nights reassembling for the next day's dive. I walk to my station and return with a cylindrical plastic tube core and plug it into a doughnut hole. It's a perfect cookie-cutter fit—with this hole, and that hole, and the next one. A few of the holes seem squeezed or folded, as if something landed on or near them and deformed the mud while it was still soft. Yet the tube cores appear to be a good fit even for these misshapen holes, though I can't actually test them.

Only Anand Patel, a graduate student from the University of Southern California, seems interested in this apparent match between tube cores and sediment holes. We discuss the fact that tube cores can sample soft sediments but not hard rock and speculate that the process of biopsying might conceivably prompt a reaction that turns the mud to stone, perhaps something like scar tissue. He's interested in the chemistry. I'm interested in the fact that all the scientists who love mud so much might be hardening its arteries here.

But it's just as possible or even likely that science's footprint in the deep will prove useful. With so many of the whale falls and tree falls gone, perhaps a new life form will evolve to feed on ballast bricks.

Before abandoning the sorting, I discover two worms I'm told are interesting—code for *they might be new species*. One is taken from me and secreted away in a vial of formaldehyde. The other I'm instructed to transfer to my sort dish for Brigitte Ebbe, from the Zoologisches Forschungsinstitut in Germany, to look at.

Strangely, horrifyingly, I lose it. Clamped between the tips of my paper-thin forceps, it's invisible to the naked eye and . . . somewhere along the four-inch pathway between examination dish and sort dish it disappears.

I realize this is one of perhaps many reasons why no one is enthusiastic about my sorting. I don't tell Brigitte, but someone else probably does. She's a member of a declining species herself, a taxonomist as well as a worldwide authority on dorvilleid worms. At a glance, she can tell one identical-looking worm from another, a useful skill nearly supplanted these days by costly and time-consuming DNA analysis.

A few days later, in the same fashion, Brigitte loses what she knows to be a new species. Just like that.

The strange chemical communities living beyond sunlight in the oases of whale falls, cold seeps, and hydrothermal vents may be more significant than even Jules Verne could have imagined. In 2000, the discovery of an entirely new kind of vent stirred the cauldron of human thought regarding the beginnings of life on Earth. The new genesis story was engendered by a discovery 2,300 miles east of Jacksonville, Florida, and 2,600 feet below the surface of the Atlantic atop a submerged mountain called the Atlantis Massif.

The vent was named Lost City for its mountaintop home and its discovery by researchers aboard the research vessel *Atlantis*. Lost City proved exceptional from the beginning, its fantastical metropolis populated by 200-foot-tall chimneys of white carbonate, the same limestone minerals found in most cave stalagmites and stalactites. Its vents spewed an altogether different brew from those of black smokers or white smokers, the types of hydrothermal vents known at the time.

Black smokers belch iron- and sulfur-rich minerals from scalding waters of up to 760 degrees Fahrenheit. White smokers emit anhydrite minerals at temperatures up to 570 degrees. But Lost City's clear fluids are poor in metals and diffuse at relatively cool temperatures of between 100 and 200 degrees. The thermal difference is a result of the vent's location about ten miles *outside* the hot, spreading center of the Mid-Atlantic Ridge. Lost City's cooler heat comes from a chemical reaction known as serpentinization, from interactions between seawater and carbon dioxide–rich mantle rock that produce hydrocarbons—the organic carbon and hydrogen building blocks of life.

Radioisotopic testing reveals that Lost City's hydrocarbons do not come from the living world of the biosphere—that is, they have not drifted down from the surface. Instead, these organic building blocks are entirely sired by inorganic parents—and prolific ones at that. Serpentinization at Lost City produces up to a hundred times more methane than a typical black smoker—and methane is to the life of the deep what sunlight is to the life of the surface: the fuel that ignites the food web. Furthermore, Lost City's hydrocarbons are even more complex than those at other seeps and vents. Its menu includes acetate, formate, hydrogen, and alkaline fluids—substances likely to have been important energy sources for the ancestors of the methane-eating microorganisms now fueling vents and seeps.

Most intriguing of all, Lost City is an extremely long-lived site. Carbon-14 dating reveals hydrothermal activity here dating back at least 30,000 years. Modeling of the vent field suggests that serpentinization may have been going on for hundreds of thousands of years—long enough for the flash of life to have first sparked into being. Such a system may have been characteristic of early hydrothermal environments, and Lost City's primary discoverers, Deborah Kelley and Jeff Karson, suggest that such places may have been key to the emergence of life on ancient Earth.

Some believe life arrived on our planet from outer space on

the back of a meteor or a comet. But Lost City hints that the ingredients of genesis were generated by geological processes right here on Earth—perhaps in the deeply sheltered womb of the abyss, in unseen chemical couplings between the gods of rock and the goddesses of seawater.

28

Nemesis

IN THE LICENTIOUS FASHION of the ancient gods, Oceanus, who flowed as a river to unite the air and water and create the *oikoumene,* married his sister, Tethys, goddess of the sea. A Titaness of the old times, she was also a child of Uranus and Gaia, and with Oceanus she produced some three thousand daughters, known collectively as the Oceanids, each a nymph protecting and enchanting a spring, river, ocean, lake, pond, marsh, cloud, pasture, flower. The Oceanids included Ephyra, for whom the immature medusa of a jellyfish is named; Eidyia, goddess of knowledge; Metis, goddess of deep thought; Calypso, island jailer of Odysseus; Philyra, goddess of writing and papermaking; Rhode, whose name was given to the rose; and Styx, spirit of the river that flowed nine times around Hades—the Greek version of the Vedic Rasā.

Also among the three thousand Oceanids was Nemesis, goddess of retribution against arrogant men and gods. She was best known from a statue chiseled by the famous sculptor Phidias, who worked an uncarved block of marble that had been confidently carried to war by an invading Persian army so sure of the outcome of the upcoming Battle of Marathon that it was ready to commemorate victory with a stele before the battle was fought. In gratitude for what turned out to be a Greek victory, Phidias carved a likeness of Nemesis from the Persians' marble.

Nemesis dogs us always, and Oceanus is remembered in many ways, but in the past two thousand years Tethys, mother goddess of the sea, has largely disappeared from human memory, her divine mantle dwindling to oblivion. She was resurrected in 1893 by an Austrian geologist named Eduard Suess—an early exponent of ecology, who coined the term *biosphere* for the wedded living realms of the atmosphere, hydrosphere, and lithosphere—who named an extinct ocean for Tethys. Later paleogeologists expanded her namesakes to include the Paleo-Tethys Ocean and the Tethys Sea—three bodies of water not so much distinct from each other as a single body shape-shifting between eons when the crustal plates of a restless planet built and rebuilt oceans and continents.

Suess's Tethys Ocean is now largely subsumed by the Indian Ocean and southern Asia. But its birth around 200 million years ago split the supercontinent of Pangaea in two, opening the way for a revolutionary redesign of the deep blue home, of Oceanus himself, his watery and airy currents, and so of Earth and the children threatening to become Earth's nemesis.

Bad weather at sea is exponentially worse than bad weather ashore. The liquid world reacts pyrotechnically to blowing air, exploding into the marine equivalent of a firestorm at winds that onshore might only make you button your coat. A tempest is brewing now, spawned by a cold front from the north tangling with two tropical systems—the remnants of 2005's Tropical Storm Tammy and Subtropical Depression Twenty-two. The confluence is birthing a midlatitude cyclonic monster destined to grow to 1,100 miles in diameter.

I've managed to catch a ride out of Woods Hole, Massachu-setts, aboard the research vessel *Oceanus,* straight into the heart of this frightful brew. A day earlier, en route to Cape Cod but stuck in Chicago by conditions bad enough to close Boston's Lo-gan Airport, I called Ruth Curry, the expedition's chief scientist, to ask what she made of the forecast. Science doesn't stop for the weather was her cheery reply.

Twenty inches of rain have already fallen over parts of New England. Coastal evacuations are underway. *Oceanus*'s thirteen-person crew, salts old and young, with lifelong careers on the water, warn us we're in for a wild ride, destined for a Force 9, a strong gale on the 12-point Beaufort scale. Before we make land-fall one week hence we'll have dabbled in Force-10 storm winds and skirted Force-11 violent storm conditions. Force 12 is a hur-ricane. Knowing the forecast, most of the eleven members of the science crew—oceanographers, chemists, technicians, and grad-uate students—scatter across the decks in the moments before sailing, seeking privacy for last phone calls home.

Outside Buzzards Bay we're slammed with twenty-foot seas ripped white by wind and careening unpredictably on the shal-low waters of the continental shelf. The swell runs abeam of us, and *Oceanus* wallows with the corkscrew motion sailors despise. One by one, those of us not on watch disappear below to set the storm rails on bunks, wedge life jackets under mattresses, climb in, wait, and hope for intestinal fortitude and good seamanship from the crew on the bridge. The only way to avoid being flung from our bunks by the violent motion is to hold on and hug the wall—the inside of the vessel's outer skin.

It's a strangely intimate experience below water line, feeling the ship bowing and flexing against our backs and absorbing into our bones the deafening thunder of steel as the largest waves drive *Oceanus* nearly to a shuddering stop before her single pro-peller fights back with the power of 3,000 horses. I'm torn between staying awake and worried in a fascinated kind of way and falling into oblivious sleep.

Concerns about weather are part of what is sending us to sea in the first place, notably concerns about that weather maker the thermohaline circulation. The goal of the expedition is to sail two thirds of the way from Cape Cod to Bermuda along a 321-mile-long transect.[1] We're scheduled to sail outbound non-stop for thirty-six hours and work our way back, sampling waters from the surface to the abyss at twenty-two predetermined stations identifiable only by their latitude and longitude. In the course of a week we'll measure temperature, oxygen, salinity, and chlorofluorocarbons in the water column of the Deep Western Boundary Current.

Ruth Curry's work aboard *Oceanus* is part of a five-year study monitoring the thermohaline circulation. Earlier in 2005 she published a paper in *Science* calculating that since 1965 some 4,500 cubic miles of fresh water from rivers and icemelt have been added to the cold waters between Labrador and northern Europe. Based on the trends of the last forty years, she estimated, it would take another hundred years of similar freshening to shut down that critical element of the thermohaline circulation, the Atlantic meridional overturning circulation, the primary heat-transport mechanism awarding northern Europe a climate more like New England than Alaska. But add enough warming, evaporation, and fresh water—speed up the cycle as has clearly happened since 2005, with the Arctic and Greenland thawing at record speeds—and there is potential for enormous change on an accelerated schedule.

The weather asserts itself from the start of the cruise, forcing Ruth to reverse the order of visits to each of the stations on the transect. She is also compelled to suspend deck operations for one critical night, when huge waves wash aboard in the dark-

ness, swamping her to her waist and knocking her off her ~~~
risk of being swept overboard. Some of her crew are down with
seasickness in the worsening conditions, and all of us are dulled
by insomnia and by exhaustion from constantly compensating
for pitch, roll, and yaw. I've combined two powerful seasickness
medicines, a strategy awarding me an hour or two in a strange
Quaalude-like realm, where I have to remind myself to breathe.
But I'm on my feet now, or rather on my backside, wedged into a
stuffed chair in *Oceanus*'s library and chuckling helplessly at car-
toons in *The Prehistory of the Far Side.*

Do you want to work? Curry prompts.

Suddenly, I'm on deck operations, geared up with hard hat,
foul-weather gear, and life vest, crouched on the starboard deck,
where unpredictable waves wash over the rail and swamp us to
our knees.

We are tending the workhorse of oceanography, a five-and-
a-half-foot-tall stainless steel contraption known as a conduc-
tivity-temperature-depth profiler, or CTD—a collection of plastic bottles arranged in a circle and mounted on a round frame.
The package, as it is known, also contains an acoustic Doppler
current profiler to record water velocity. At each stop, the 1,400-
pound rig is launched overboard with a winch and lowered to
the bottom to capture water samples and other data along the
way.

Launching and retrieving requires a hydrographic boom and
30,000 feet of coaxial cable. Performing these operations in heavy
seas requires skill and coordination from the bridge, the winch
operator, the bosun, and the science crew manning the gaffs and
lines. Using only a single screw and a bow thruster, the bridge
crew must hold the ship steady in swells of twenty feet and more,
while all others involved must ensure that the streaming cable—as powerful as a band saw—never contacts *Oceanus*'s steel
hull.

Adding to the perils of the atrocious weather, as the CTD descends it enters a series of water masses of different density gradients. These are the underwater rivers of the thermohaline circulation, each powerfully flowing in its own direction and with its own velocity—a reality made obvious topside when the cable suddenly whips through the water as if hooked to a giant fighting fish.

Ruth calls it blue-collar oceanography. The basics of it—big ships, global navigation satellite systems, depth finders, gyrocompasses, winches, cranes, and miles of cable—are the stuff of modern scientific seafaring. Yet in one way nothing has changed. Working the ocean still requires a delicate finesse between audacity and deference.

Beyond the continental shelf, the seas stretch out in the deeper waters of the Gulf Stream, easing our ride slightly. We emerge from our bunks, from stations in the dry lab, wet lab, library, galley, and engine room. Science crew, ship's crew, everyone who has a free moment abandons the interior for a break in the air and the weather. This is a ship full of seafarers who love the deep blue home.

On deck we are offered the most glorious of ocean sights, unobstructed horizons and armies of oversized swells marching abreast, topped with steel-gray helmets and flouncy feathers of foam. Albatrosses skim wind off the wave fronts, riding the long scythes of their wings like boomerangs destined to circle an ocean before returning to their sender. Whales voyage here too, the secretive deep-sea diving species of the oceanic zone, rarely seen beaked whales and bottlenose whales. But the wind knocks down their blows before they arise, fortifying their mysteries.

At night the glories offshore amplify and deepen. Far from land and all its light pollution, the evenings produce epic sunsets veiled behind screens and scrims of cloud, each layer lit with a private rainbow restrained within the burnished spectrum of blood and brass and edged with twilight and a true navy blue. Between scudding waves of stratocumulus clouds lie fields of stars and a moon waning toward its final quarter.

A few of us venture onto the upper weather decks at night, taking shelter from the wind and from each other, seeking silent communion with the forces of climate, wind, cosmos, sea, and darkness, aboard what is, after all, a frail vessel—as are we. These are the moments that bind us not only to the mysteries of our universe and the mysteries of ourselves but to all sailors who have gone before, stripped of their own technologies, whatever their time, bared before the hard edges of a world that forces us to its whims. Rán is at work out here.

At night the bioluminescence of the sea runs bright. It's an aquatic aurora, formed by phytoplankton that flash in response to the motion of the waves and everything swimming through the waves. Much of the food web of the sea consumes the phytoplankton or consumes those who consume the phytoplankton, and many of them absorb the light-producing pigments of their prey, so that much of the ocean's phosphorescent glitter is a gustatory side effect . . . as if you and I could grow flight feathers by eating chickens.

Oceanus's wake sparkles and pops in dazzling ephemeral flashes, the eddies and swirls scintillating behind us like a blue brick road, marking our path up and down the humps of the swells. Where the waves break at the top, the luminescence shatters explosively, as if Ægir were detonating fireworks in his great hall beneath the sea. *Oceanus*'s hull, enveloped in the glow, shines like a spaceship against the stars in the water and the stars in the sky.

At one station, as the ship fins and paddles with micromovements of bow thruster and aft screw, as the steel cable feeds out the CTD like a hard-running marlin, I huddle on a windward deck, feeling the full force of the dark and the weather. The night air is full of the humid organic breath of the tropics and also of the cold sheer of Hræsvelg from the north, the scents battling each other with each riposte of wind.

Torpedoes are firing below, hundreds of sleek aqua-green bullets curlicuing crazily through the sea. These are the signatures of squid drawn to the deck lights of ships, the fastest and boldest pens, veritable John Hancocks of the deep blue home. There are far too many to count—better to just let their names write and dissolve without thought. And then, lumbering up from within the group, on the slowest and steadiest of cursives, like a barnacle-encrusted shipwreck returning to the surface, comes the capsized dinghy of a leatherback sea turtle, enveloped in an aura of fizzing, chattering light of every ethereal shade of fairy blue and nymph green.

She is in no hurry compared to the squid, although it's likely that out here beyond the continental shelf she is returning from the jellyfish-rich realm of the mesopelagic zone, perhaps from a dive to 4,000 feet. She has certainly passed the CTD going the other way, an anomaly in a universe of anomalies below the waves. Perhaps she is drawn to the riddle. Perhaps she is following squid. At the surface, her head bursts through the bubble of bioluminescence. Haloed by deck lights, she tilts her head to drink the air from dueling storms.

The work continues day and night. Like the squid, we run an erratic course, dodging the low-pressure systems that shape-shift around us on an hourly basis. We miss a few transect stations completely, chased off by weather. But we are happy to persevere, and we succeed at most.

Toward the end of the week, surrounded on every horizon by menacing black skies, complete with downpours and bolts of lightning, we bask for an hour or two in a spotlight of sunshine that illuminates the endless cobalt of the deep, the platinum spray of the surface. Three of us—Ruth Curry, Guy Mathieu, and I—are out on deck tending the CTD, which has just returned from its four-hour journey to the bottom of the ocean. Mathieu, a retired research associate with the Lamont-Doherty Earth Observatory, is collecting samples from the Niskin bottles for analysis of their chlorofluorocarbons, the synthetic chemicals in refrigerants and aerosols that are so damaging to Earth's stratospheric ozone yet so useful as tracers for measuring the time scale of movements within the ocean conveyer belt.

Ruth taps the bottles for oxygen analysis, and I follow up with salinity samples. Although conditions are wet, rough, and slippery, we smile, enjoying our time on deck. Five hundred miles from land, we are deep inside the embrace of the ocean, and as we work we are touching water that an hour or two ago rode the Deep Western Boundary Current 17,000 feet below us, headed for Antarctica. The sea, always a place of awe, is made even more so by the feel of its cold buried tides.

Although we do not yet know it, the thirty-year trend of freshening in the North Atlantic is beginning, paradoxically, to reverse, as the high latitudes are flooded with warm salty waters from the subtropics. It's part of a long-term natural cycle and part of what makes finding the embedded signature of anthropogenic change all the more challenging, since the natural cycle is scrambled alongside human-induced changes. Later Ruth will tell me that greenhouse warming is driving the oceanic system to greater extremes, with higher high salinities and lower low salinities. Because of us, Ægir's signature brew is changing.

But at this moment we are happy to be at sea in bad conditions collecting good data that may well lead to bad news. The tempest around us is beautiful yet seemingly manageable—that is, until the winds, whistling steadily at 40 knots, increase sharply, ripping the whole surface off the sea, not just the tops of the swells.

One of Ægir's daughters, Kára the Powerful, has appeared. The whistling grows ominously louder and splits into harmonics of deeper- and higher-pitched voices. Literally over our heads, the low-pressure storm systems have merged, and within the hour we are running south as fast as *Oceanus* will go.

No one who survives time at sea is ever less than humbled by its powers over life.

29

The Inexplicable Waves

IN THE VIKING CREATION STORY, before there was a universe there was Élivágar, the inexplicable waves running through the Ginnunga Gap, the immeasurable void of potentiality. From the waters of Élivágar emerged the first life, the first *jötunn* of the race of giants, or elder nature deities. His name was Ymir, and his appearance put in play the forces that would eventually lead to the birth of the gods, who would dismember him to make the cosmos, turning his bones into mountains, his flesh into earth, his blood into ocean, his skull into heaven.

Both *jötnar* and gods knew that at some point in the future the world would end in the course of a mighty battle between their two kinds. One giant in this as-yet-unplayed war was Angrboda, a *jötunn* of Ægir's time. Her name meant She Who Offers Sorrow—perhaps a reflection of the fate that befell her three children.

Angrboda's offspring terrified the gods, and eventually all three of her strange brood were cast from the world by Odin, god of gods, and his sons. They bound Fenrir, the Wolf Child, with a silken ribbon made by dwarfs from six secret ingredients that consequently no longer exist on Earth: the sound of a cat's feet, a woman's beard, a mountain's roots, a bear's sinews, the breath of fish, and bird spit. They cast Hel to the underworld, where she ruled the dead in a great hall called Rain-damp, built over a threshold called Stumbling Block. The gods threw Jörmungandr, the World Serpent, into the ocean, where she grew large enough to encircle the land and bite her own tail.

Odin was aware that he would be killed by Fenrir and that the world would end in the great battle of Ragnarøkkr, the Twilight of the Gods. Yet despite his foreknowledge, apocalypse came anyway. Odin was swallowed by Fenrir and immediately avenged by his son, Vídar, who tore the wolf apart at the gullet. Thor battled Jörmungandr, and the great serpent's death throes spun the waters of the world into chaos.

The end of the battle of Ragnarøkkr marked the death of the *jötnar*, the death of the gods, the end of the world as it had been since before time began, since the beginningless onset of the inexplicable waves. A few younger gods survived to begin a new world, as did Hel.

Although scientists do not speak of it, we may: the impending threats to Earth are as monumental to our world as the battle of Ragnarøkkr was to the ancient Norse. Chief among these threats are the enmeshed entanglements between global climate change and the sixth great extinction—each a symptom of the other, each a cause of the other, each perhaps more a matter of the ocean than of the land or air.

In a sense, these entanglements are the *jötnar* and the gods still afoot in our world. Scientists have identified twelve by name—twelve global-warming tipping points delicately aligned between ocean, atmosphere, land, and life, planetary levers encompassing powerfully diverse yet interrelated systems: the Asian monsoon, the Amazon rainforest, the oceanic currents of the North Atlantic and the Antarctic, the ice sheets of Greenland and Antarctica, the Sahara Desert, the Tibetan Plateau, the ozone hole, frozen deposits of methane, salinity valves in the oceans, and El Niño. Tipping any of these systems beyond a precisely interlocked stasis could initiate sudden catastrophic changes across the globe as others cascade alongside.[1]

Many linkages exist to stabilize these tempestuous *jötnar* and angry gods, some unassuming, others invisible. Among the strongest linkages are the living ones, built double-helix tough. Chief among the living are the near omnipresent phytoplankton, the humble microscopic plants that, in the course of photosynthesis, remediate a hefty dose of our atmospheric carbon dioxide emissions by sequestering them underwater. All told, the ocean has mitigated some 85 percent of the excess heating of our atmosphere since the 1950s—though not without a cost.

Increasing levels of carbon dioxide are incrementally acidifying the World Ocean, and experiments now suggest that the shells and skeletons of everything from phytoplankton to reef-building corals will begin to dissolve within forty-eight hours of exposure to the acidity expected in the seas by the year 2100. One forecast predicts that the rising levels of atmospheric carbon dioxide will condemn coral reefs—the seas' most biodiverse realms and food sources for one in six humans—to extinction within fifty years. Jörmungandr, the World Serpent, is chewing the bones of her tail, one vertebra, one seashell at a time.

The threats to all life, including human life, from changing plankton, a changing climate, and a changing ocean far exceed even the calamities looming from rearranged icecaps and coastlines, swelling disease boundaries, and shrinking agricultural zones. Phytoplankton perform the most complex process of the universe: creating life from nonlife, using the energy of sunlight to synthesize food from carbon dioxide. They perform this transcendent mission countless times every minute of every day to construct the living aura of our planet, one microscopic particle at a time. The combined result of their separate labors is to build the foundation of the deep blue home, a food web supporting huge biomass on land and sea, feeding the living Earth, including you and me.

Nor is that the end of the vital services performed by phytoplankton. The coccolithophorids—spherical cells less than 100 micrometers in diameter and armored by calcareous plates resembling microscopic Life Savers—are significantly responsible for the formation of clouds in our atmosphere. Their dimethyl sulfide emissions convert to sulfate molecules in the air, becoming the nuclei for cloud condensation and cloud cover. Thus the tiny plants of the sea account for much of the rain falling on the innermost reaches of the land.

The tiny powerhouses that are plankton also play a critical role in keeping the World Ocean circulating. The oceanographer William Dewar of Florida State University calculates that the marine biosphere, starting at its broad base, the plankton, invests about one terawatt in the mechanical energy of swimming, effectively mixing the World Ocean from within. Amazingly, the swimmers of the deep blue home provide a third of the power required to blend the waters of the world from the abyss to the surface, with the remaining two thirds powered by the more obvious forces of wind and tides.[2] So it is that life itself helps rotate the great waterwheel of temperature and salinity driving the underwater rivers of the world and making our planet habitable.

Humans are a terrestrial species biased toward attributing the forces we see around us to familiar forces on land. But the more we look, the more we learn that everything arises from the sea and everything falls away to the sea, and the deep blue home is home to every one of us, whether we are beings of the water, air, rock, ice, or soil.

30

At the End of Hunger

ACCORDING TO THE ROMAN POET OVID, Nemesis exacted her harshest retribution against the lovely youth Narcissus, who was conceived during the ravishment of the blue nymph Liriope, after her encirclement within the oxbows of the river god Cephisus. Narcissus was born so beautiful that his mother worried. How long could he live like that? She consulted the prophet Teiresias.

If ever he knows himself, he will surely die, was the reply.

By the age of sixteen, Narcissus was so indescribably exquisite that all who set eyes on him loved him, male and female, young and old. Yet he scorned all. No one was comely enough for him.

One day, hunting stags, Narcissus was spied by Echo, a mountain nymph, who fell instantly and irrevocably in love. But she was afflicted with a unique punishment brought down upon her by Hera, wife of Zeus. For too long Echo had entertained Hera with fantastical tales, keeping her in rapt attentiveness and allowing Zeus the freedom to cavort with other mountain nymphs. Hera eventually realized the ruse and silenced Echo's voice except as a mimic of others'.

Thus, when Echo met Narcissus, she was blighted with love but unable to say anything except to return his words: Who's there? *Who's there?* She followed him but could not speak her own heart, and he repulsed her. In time she moved away and in her suffering hid alone in the darkness of woods and solitary caves until her life force was spent and only her voice remained.

Narcissus continued to attract forlorn lovers like flies to an inaccessible feast, and still he would not return one morsel of their love—until the day another virgin fell in love with him and was rejected. In her misery she prayed that Narcissus should suffer unrequited love too, and Nemesis, goddess of retribution, heard the plea.

In short order the boy saw his own face in a forest pool and fell as lethally in love with himself as had everyone else. Like all the others, he could not leave this beautiful youth, and time and again he leaned down to kiss the perfect smiling lips, reached out to embrace the perfect reaching arms. But the boy he loved was impossibly elusive. Eventually Narcissus realized that his loved one was himself and that his love would kill him:

> *As wax dissolves, as ice begins to run,*
> *And trickle into drops before the sun;*
> *So melts the youth, and languishes away,*
> *His beauty withers, and his limbs decay;*
> *And none of those attractive charms remain,*
> *To which the slighted Echo su'd in vain.*[1]

The Norse held on until about 1450 in Greenland, not much longer than the Dorsets/Tuniits who had sailed to their seal hunts atop inflatable sealskin life preservers. By 1450 the weather had grown appallingly colder and the ice had grown obstructively back across the seaways—just as the Dorsets would have liked it, had they survived.

The poet Al Purdy, imagining the end of their world, questioned whether the Dorsets realized what was coming their way, whether an old hunter, carving a figure retrieved from an image in his mind and bearing its weight through his arm into ivory, ever saw his own future.

> *The carving is laid aside*
> *in beginning darkness*
> *at the end of hunger*
> *and after a while wind*
> *blows down the tent and snow*
> *begins to cover him.*

The changing climate in the late Middle Ages—delivered by what may well have been the changing circulation of the ocean—fatally undermined the hay-growing season and therefore the Vikings' way of life in Greenland. Whether the last of them sailed back to Iceland or Norway or died in Greenland is unknown—although, as Robert McGhee writes in *The Last Imaginary Place*, an Icelandic sailor named Jon Greenlander, driven to Greenland by storms about 1540, found abandoned farms and boat sheds like those of Iceland. "And when they landed he found a corpse clothed in a mix of woolen cloth and sealskin, carrying a knife with an iron blade that had been honed to a sliver." McGhee, a man with many Arctic decades under his belt, imagines the end of the Norse in Greenland:

They were leaving a land where their ancestors had worked and built a European way of life that had survived as long in Greenland as European-based societies have now lived in the Americas . . . Occasional small families must have succumbed to hunger in the winter darkness of unheated houses, or died where they worked on the beach, like the man whose corpse Jon Greenlander reported.

The Little Ice Age that put an end to the Norse residency of North America may have resulted from melting Arctic ice and the faltering of North Atlantic currents in the wake of a few centuries of warm weather. But new evidence suggests that the cooling was abetted by a powerful though distant collaborator initiated by none other than Christopher Columbus.

Prior to Columbus's 1492 voyage, somewhere between 10 million and 100 million people inhabited the Americas. At least 80, and perhaps 90, percent of them died in the decades following the arrivals of Columbus, Pedro Álvares Cabral, John Cabot, Ferdinand Magellan, Giovanni da Verrazano, Jacques Cartier, and others. The ancient native civilizations of what Europe called the New World disappeared virtually overnight from epidemics and from genocides, taking with them their agricultural endeavors, both farmlands and wild croplands, including fields of native maize, potatoes, beans, cacao, vanilla, tomatoes, chilies, cranberries, squash, pumpkins, pecans, wild rice, sweet potatoes, blueberries, and peanuts.

A study by the Stanford University geologists Richard Nevle and Dennis Bird assessed the amounts of charcoal in soils and lake sediments at pre-Columbian sites and compared these data to the amounts at post-Columbian sites. They concluded that the massive natural reforestation after the abandonment of countless millions of acres of agricultural lands pulled so much carbon out of the atmosphere as to change the climate—enough, they calculated, to have significantly contributed to the Little Ice Age.

And so it may be that the Dorsets died and the Vikings fled in part because thousands of miles to the south, the Aztecs and the Incas died, and the tendrils of new trees reclaimed their cornfields, reaching into the air and into the ground, redrawing the carbon cycle and therefore the chronicles of fish, birds, whales, people, forests, vines, farms, temperatures, currents, ice, rain, snow, glaciers, bones.

• • •

McGhee theorizes that some of the last Norse Greenlanders did not entirely quit the New World but joined the men working a new cod fishery emerging in Newfoundland about the time they abandoned Greenland. This fishery remained a seasonal enterprise without permanent settlement until about 1610, when men from the Viking coast of southeast Ireland came to live on the island of Newfoundland, naming it Talamh an Éisc in Gaelic, *land of the fish*. The fish they spoke of were cod, of course, those bottom dwellers who reached more than six feet in length and weighed hundreds of pounds, who were so abundant, so easy to catch, that even in the course of the short boreal summer, enough could be hauled ashore, cut into strips, and dried in the foggy wind to sustain a family through an ice-locked Newfoundland winter.

These settlers encountered the native people of the island of Newfoundland, the Beothuk, whose language may have been related to the Algonkian languages of the Mi'kmaq, Blackfoot, Cheyenne, Yurok, and many other groups across Canada and the northern United States. We know little of the Beothuk people, although archaeological excavations near Lewisport on the island of Newfoundland have revealed large oval houses the Beothuk called *mamateek* and burial sites that included small carvings of bone, flattened like pendants and rubbed with red ochre. No one today knows what any of these represented.

The last known surviving Beothuk was a young woman named Shanawdithit, who was born about 1801 on the edge of a lake somewhere on the island of Newfoundland. This might have been an unusual birth site for her people under normal circumstances, but by that time European fishermen had driven the Beothuk into the interior, where they were persecuted by trappers and furriers.

In 1827 Bishop John Inglis of Nova Scotia visited Newfoundland and had occasion to meet Shanawdithit, hear something of her story, and visit some of the places of her people. He relayed what he knew in letters and reports, describing her land as finely wooded with large timber, the scenery rich and picturesque, a good place, he concluded, where the abandoned remains of her people's campsites could still be found easily. He acknowledged that ever since Cabot first landed and carried off three of her unhappy tribe, the Beothuk had every reason to lament the discovery of their island by Europeans:

English and French, and Mic-Macs and mountaineers, and Labradors and Esquimaux shoot at the Beothick as they shoot at the deer . . . and none have been seen for several years, it is feared by some that a young woman who was brought in about four years ago, and is now living in Mr. Peyton's family, is the only survivor of her tribe.

I had some conversation with Shanawdithit . . . [and learned that in] April, 1823, a party of furriers in the neighborhood of the Exploits River followed the traces of some Red Indians, until they came to a wigwam or hut, from whence an Indian [man] had just gone, and near it they found an old woman, so infirm that she could not escape. They took her to Mr. Peyton's, where she was kindly treated, and laden with presents. After a few days she was left at her wigwam, while the furriers searched for others. Two females were soon discovered . . . Though much alarmed, they were made to understand by signs that the old woman, who was their mother, was at hand. The man who had been first seen was their father, who was drowned by falling through the ice [in trying to escape]. The women were in such lamentable want of food that they were easily induced to go to Mr. Peyton's . . . One of the young women, who had suffered some time from a pulmonary complaint, died as soon as she was landed . . . and very soon after they arrived there the old woman also died, and Mr. Peyton has retained her daughter, Shanawdithit, in his family ever since.

She is now twenty-three years old, very interesting, rather graceful, and of a good disposition; her countenance mild, and her voice soft and harmonious. Sometimes a little sullenness appears, and an anxiety to wander, when she will pass twenty-four hours in the woods, and return; but this seldom occurs. She is fearful that her race has died from want of food.[2]

Shanawdithit survived only another two years, succumbing in 1829 to the same tuberculosis that had killed her mother and sister. After death, her skull was separated from her remains and presented to the Royal College of Physicians in London, later to the Royal College of Surgeons, and eventually destroyed during the German bombing of London in World War II. Her headless body was buried in St. John's, Newfoundland, in a graveyard that was later dismantled for the construction of a railway.

According to Bishop John Inglis, all that could be discovered of the religion of the lost Beothuk people of Newfoundland was that they feared some powerful monster appearing from the sea to punish the wicked.

Aboard *Ceres* in Trinity Bay, Newfoundland, in 1984, our enthusiastic host, Peter Beamish, is suffering from a dampening of his natural passion. He has not been able to find any humpback whales for us to film. We don't hold him accountable. But it's always this way with our oceangoing guides. They feel personally responsible for satisfying our hopes and revealing to us all those beings and phenomena they know exist in their corner of the deep blue home. We know that *they know* this is not possible—that only Ægir is in control. But still.

Disheartened, his Tinker Bell energy dimming, Peter sails his little boat into the loins of a green fjord, plodding through the blond kelp bound to the seashore, skirting the submerged rocks and grounded icebergs. He is taking us to see the clattered bones of a ghost village sagging to driftwood up on the hill. It lies on a small island known as Ireland's Eye—named for another Ireland's Eye, a tiny island clear across the Atlantic off the coast of

Dublin. The original Eye was known in Celtic times as Eria's Island. But the Vikings who settled that Irish coast substituted their word for island, Ey, and Eria later morphed into Éireann, Gaelic for Ireland.

A thousand years after the Vikings and two thousand miles west of the first Ireland's Eye, Peter leads us up a hill slippery with new grass. The ghost village is impossibly picturesque, nestled in the intimate embrace of a tiny harbor edged with tuckamore and rock. Some of the planked, two-story, no-nonsense houses of this erstwhile fishing village still bear splashes of turquoise, mustard, maroon, and fire-engine red on their leeward sides and under their eaves—the colors of painted houses even now in Newfoundland, Ireland, Iceland, and Norway. A few buildings have collapsed completely to skeletal planks. The others are slowly settling into a landscape scraped to the rock by glaciers and barely endowed with soil ten thousand years past the retreat of the ice. All are reflected in the still waters of the inlet.

It's the season of bloom, however brief, and the grass between the ice-polished rocks is enlivened with blossoms of wild madder, yellow pea, pearly everlasting, pink touch-me-not. We trudge with our cameras to the top of the hill, along pathways still worn into the boreal soil by feet that have not trod here regularly since 1949, when the government of Canada withdrew the tentacles of its support for Newfoundland's outport fishing villages accessible only by sea.

An old collapsing church stands atop the hill, positioned to sight straight down the seaway, through the embrace of the little protective harbor and beyond—out into the vast vault of the North Atlantic. Hræsvelg's wind is gathering up here, and we watch over the water as twelve bald eagles (*Haliaeetus leucocephalus*; Red List: Least Concern 2008) pair up to perform aerial acrobatics, the pairs flipping in complete circles as they meet, talons to talons. One pair passes what looks like a wooden plank back and forth.

Within the purview of the village lie the remains of two grave-yards, the old one and the new one, though both record the same handful of surnames: Cooper, Loder, Hodden, Kelley, Watton. Directly behind the old church, tucked into the tuckamore, stands a lone gravestone, that of one James Toop, beloved husband of Sarah Toop, who died at the age of seventy-nine in May 1890. Here was a man who shared this island with Shanawdithit for the first eighteen years of his life.

The old cemetery holds some fifty graves, including that—or at least the grave marker—of Maxwell George Cooper, who drowned on June 29, 1911, aged three years and eleven months. His inscription reads:

> *He sent from on high to fetch me*
> *and took me out of many waters.*

The same could be said of all who live here, all who live and die in Newfoundland, Ireland, Iceland, Norway, or any other shoreline of the deep blue home. "He" could be Ægir or Oceanus or Rasā or Rán or Yahweh or Muhammad or Jesus or Sedna or a fearful monster from the sea whose name we have forgotten.

Somehow the small inscription in the lonely graveyard in this ghost village on an abandoned island cheers Peter. He shakes off his gloom, regathers his enthusiasm around himself like a cloak of wildflowers, and skips down the hill. We can hardly keep up as he leads us back to *Ceres*. He believes once again in the possibility of a humpback whale chasing a shoal of capelin fleeing their nemeses the cod, and of that humpback whale finding herself entrapped in a fisherman's net, and of both net and whale needing rescue, and of there being a job at last for a hero in this heroless age. He sings as he bounces toward the sea, and his voice echoes from the rock around us.

PART THREE

~

THE AIRBORNE OCEAN

The moth and the fish-eggs are in their place,
The bright suns I see and the dark suns I cannot see are in their place,
The palpable is in its place and the impalpable is in its place.

—WALT WHITMAN
Song of Myself

31

Serpent Cave

WE ARE LOCKED in the embrace of the high desert, held motionless by scale—by the immensity of the mountains and a sky so clear we peer through it, thinking to glimpse the other side. A hard wind blows from the north. We list in our saddles, compensating. Much is in flight on the wind: the petals of orange poppies, bumblebees bouncing between the refugia of yellow cactus flowers, butterflies. We have no shelter, except in the haven of observation and in our leisurely pace—time enough to witness the ways the long eons have worn the youth from this landscape, worked its bones through at the joints, labored its muscles to sinew.

Our course is relentlessly uphill and downhill. Up weatherbeaten plateaus thousands of feet high just to descend spidery trails into sunburnt canyons hundreds of feet deep. There is no other sign of human life here and certainly not any sound of it, no planes, cars, or voices. The silence is so thick our ears feel congested, and we jump at sparse noises that startle us with their sudden clarity—musical donkey bells, a wren's soliloquy, clickety grasshoppers, doves.

Here in the high desert mountains of Mexico's Sierra de San Francisco, far from clouds, rivers, ocean, the landscape swells with bloom. White palo blanco trees shade themselves with leafy mantillas. Portly barrel cactuses strain the waistlines of their accordion suspenders. Ocotillos crook red flower tips from whippy arms. In every crack of the red and yellow rock of this world, green gleams as bright as water.

Only a few springs and secret water holes enliven this range. More importantly, ancient trails woven through topographies of air deliver sustenance from the Gulf of California, forty miles to the east, and the Pacific Ocean, one hundred miles to the west. Powerful conveyers progress constantly across this dry landscape, wave after wave of vortices, drifts, turbulence. Sometimes the fluid dynamics corporealize in the balletic choreography of clouds. But even in clear skies the flow of damp swirls invisibly.

All life here focuses on the capture and storage of the airborne ocean. Many animal species retreat underground, into a tangled overexuberance of canopylike plant roots, into an underground desert as lush as a jungle. Animal burrows crisscross this subterranean forest—private toll roads engineered, inhabited, stolen, shared, and fought over by ants, beetles, bees, wasps, cicadas, tarantulas, spiders, lizards, snakes, mice, rats, foxes, tortoises, and badgers. Coyotes dig and maintain coyote wells, which many other animals and plants avail themselves of. Cactus doves transform their bodies into flying canteens, imbibing 15 percent of their body weight whenever water becomes available. Box turtles store urine in oversized bladders and urinate on themselves to stay cool. Mesquite trees grow taproots 160 feet deep to mine perennial moisture. Every particle of wet that falls from above arises again in the living bodies of the dry world.

There was snow here last winter, explains Juan Carlos Arce Arce. And it rained three weeks ago.

We absorb this information. Last winter was a year ago, and three weeks ago marked the onset of *this* winter. From so little, life rekindles.

Juan Carlos is a young blue-eyed vaquero and our guide into Baja California's remote Sierra de San Francisco. He is probably a descendant of an Englishman, Juan de Arce, who came to these parts in 1698 and presumably seeded the blue eyes that are common here. Juan Carlos Arce Arce was born and raised at the gates of this volcanic range and is accustomed to its sharp edges. Despite the cold and the wind, he rides with his denim jacket open, one hand on the reins, the other on his hip, hat cinched low. Enriqueta and I share a smile. We could hardly imagine a more winsome cowboy to deliver us into a world we have dreamt about for more than two decades.

It's 2001, twenty-one years past our shared field season on Isla Rasa, when we stared at the mountains to the west of our island, imagining their unforgettably secret past. Enriqueta has returned to Rasa every spring since to live among the birds. She has become one of the foremost voices for the preservation of the wild islands of the Gulf of California. Her presence, her deepening knowledge, have protected this corner of the world far beyond the modest funding she receives.

She has also become an accomplished horsewoman, an endurance rider who sits her mule easily. I have not ridden since teenagerhood and am feeling keenly the firmness of my saddle and the leniency of my will—unable to control my mule, a gelded sweetheart named Rocío (dew). He is a gourmand of the desert and particularly fond of biznaga cactus, pausing at every one along the trail, stretching his big head to swipe sideways with his teeth, lips retracted. Spines notwithstanding, he eagerly excavates the cactus in search of its succulent orange meat. Nor can he resist the tubular purple flowers and serrated leaves of yerba santa, which locals suck to quench their thirst.

Enriqueta and Juan Carlos are tolerant of my tolerance for Rocío's appetites. But now, hours into the ride, they scold me and tell me to keep him moving—though neither Rocío nor I worry that I will succeed.

And so we have time for observation beyond even the normal pace of an unhurried mule walking uphill. We have as much time as it takes for a mule with a love of food to taste his way up a verdant trail lined with juicy orange hearts guarded by thorns.

We are following Juan Carlos on his well-behaved mule, and he is following his three burros, all heavily laden with crates of food and camping gear. Despite their burdens, the donkeys are inclined to travel fast. The lead animal, a jenny, wears a bell, though she climbs so nimbly on dainty unshod feet, even on the steepest and most precipitous trails, that the melancholy chime sounds only occasionally. The burros are pressing ahead and Juan Carlos is quietly talking them back. Yet they push on, eyes cast back, thick eyelashes blinking coyly, awaiting even a flicker of distraction from their vaquero, at which they burst into a run, bell cheering loudly as they career off the trail into the spiny landscape, stampeding and braying, joyfully willing to carry the burdens on their backs forever if only they can be free.

Swearing quietly, Juan Carlos spurs his mule after them. Though he and his steed are protected from the barbed world by home-tooled leather chaps affixed to their legs, the burros run faster. The chases persist over a surprising distance, mostly downhill, birds flushing into the air, cactus skeletons cracking. Rocío takes advantage of each commotion to halt the weary business of walking in favor of the vital business of grazing, and by the time Juan Carlos has rounded up the wayward pack animals and herded them back to the trail, Enriqueta and her mule are far ahead and Rocío's happy face is wearing a beard of thorns.

After a few such pursuits, Juan Carlos dismounts to thread a rope between the burros' packs, binding them to each other. Nevertheless they run off again, bell pealing, packs clanking.

Between extemporaneous rodeo shows, our awareness swings on a pendulum, from the extreme beauty of the landscape to the extreme difficulty of the landscape. The beauty includes small herds of half-wild burros roaming these peaks, their long, saintly faces peering shyly from behind pitaya cactuses. The loveliness includes the herds of free-range goats, who, unlike the burros, do not freeze into stillness at the sight of us, hoping we will not capture them, but bawl forlornly and follow us for miles.

The difficulties of the landscape include the steep track switchbacking down nearly perpendicular slopes. In anticipation of the long downhill into deep arroyos carved by once abundant rainfall on this volcanic plateau, Juan Carlos slides our saddles back and instructs us to lean back. Descending, the mules are forced to step heavily down on their front legs, grunting with effort, and then somehow to find the room to hop their hind legs down without catapulting us over the cliffs. They manage this acrobatic feat repeatedly and gracefully, pivoting their haunches on ground no bigger than a cowboy hat, on slopes so precipitous I squeeze my eyes closed in moments of small terror. Rocío, however, is sure-footed and unflappable, leaning down while pivoting to rip at the same cactus holding the trail together.

The wind blows hot in the bottom of the canyon, rustling the leaflets of palo blanco anchored in waterless riverbeds amid jumbles of flood-tossed boulders. We ride in single file in silence, absorbing the sudden windfall of birdsong, the rolling mimicry of thrashers, buzzing gnatcatchers, slurring whistles of cardinals, chuckling ravens. We amble on the soft footfall of our mules in and out of pools of shade, twisting through dead-end trails that become blind turns squeezed beneath thousand-foot walls stepping down through layers of red basalt and yellow tuff carved by cascades of extinct rain.

Juan Carlos dismounts, and we leave the mules and donkeys to climb on foot to a shallow, almond-shaped cavern embedded, eyelike, in the canyon wall thirty feet above the floor of the arroyo. Inside this rockshelter—distinguished from a cave by its dimensions, wider than deep, the opposite of a cave—a vaulted ceiling slopes back to its dark meeting point with the floor. We wait inside, heads craned up, until our eyes adjust. When they do, the past steps forward in the form of a twenty-five-foot-long mural—a cave painting of an undulating red-and-black serpent with deerlike ears, antlers, a forked tail. The painted snake supports more than fifty figures dangling charmlike from its outlines: leaping hares, a floating seal, a swaying dolphin, dozens of human shapes, elongated and boxy, half black, half red, arms stretched overhead, fingers spread, some wearing long horns.

We lie on our backs, allowing the motion of this image to sidewind through our minds, to carry us to another world, created by an ancient people, whose vision of reality combined rabbits and dolphins in a universe energized by the waves of a snake. The pigments are vivid, bursting through their multicolored canvas of agglomerate rock. The lines of composition are deft, repeated time and again by a practiced hand. We have no doubt, lying on our backs in the cool silence of the cave, that we are being offered an utmost gift: a snapshot of the enigmatic world of a desert peninsula plunging through ocean waters at a time in the past when people drank their genesis stories and ate their gods and played out their sibling rivalries with their brother deer and sister whales.

32

Black Mirror

THE CUEVA DE LA SERPIENTE is one of more than four hundred grottos of cave art secreted in the nearly inaccessible wilderness along a three-hundred-mile swath of mountains spanning central Baja California. The combined works comprise one of Earth's greatest collections of prehistoric art. Nearly all are located far from present-day human habitations, in remote canyons and mesas, tucked into arid arroyos, up unapproachable walls, beneath crumbling cliffs. They have been remarkably well preserved from time by a bastion of drought as impregnable as a fortress.

Yet this landlocked world holds a crucial key to the fluid worlds. Inside these chapels of prehistory, ocean and sky mingle with boundless desert. Painted tuna, sardines, dolphins, whales, sea turtles, sea lions, manta rays, and octopuses cavort with pronghorn antelope, bighorn sheep, deer, coyotes, mountain lions, lynx, rabbits, eagles, pelicans. The people who painted them straddled all worlds with equal awe. They saw no need to segregate the influence and power of the solid from the influence and power of the fluid. Even more compelling, they saw no need to segregate themselves.

Few people alive today have seen the hidden rock art of Baja California. Relatively little scientific study has been made of the sites—although 250 painted caverns have been preserved inside a United Nations World Heritage Site since 1993. As best we know, the art harks from between 3,100 and 1,300 years ago, although the pigments in one rockshelter, the Cueva del Ratón, have been dated to some 5,000 years before the present. The rock art of Baja California, unlike other rock art in the American Southwest, contains no depictions of the arrival of the Spaniards—no horses, ships, or guns—suggesting that no one was painting here after the sixteenth century, and perhaps long before that.

Eighteenth-century missionaries believed that the local Cochimí people were descendants of the cave painters. Like the painted figures, their shamans wore elaborate headdresses woven into their hair. As in the paintings, the Cochimí saluted the sun with their hands raised. But the Cochimí themselves adamantly denied any relationship to the painters, believing that the artists were a vanished race of giants who came down from the north and eventually died from warring among themselves and at the hands of the Cochimí, who could not, in their own words, tolerate such strange residents in their lands.

Juan Carlos sits in the mouth of the Cueva de la Serpiente, whistling softly, whittling a piece of wood. When we are ready, when we have absorbed the resonant power of the serpent, he leads us across the arroyo to another rockshelter, more exposed and less well preserved. Its arched ceiling is crumbling into rubble. Pigment-stained boulders and pebbles line the floor. We turn them over, looking for images.

Only fragments of paintings remain on the ceiling, indecipherable parts of figures that once spoke volumes in this quiet place. The single clear picture, impossibly out of place in this waterless world, is the hydrodynamic form of a sea turtle swimming across a stratum of ashfall from an ancient volcanic eruption. She is flippering hard, cinching her lost realms forward toward the present.

We pitch camp on the floor of the arroyo among boulders fallen from above. The sun washes the walls of the canyon, the light arcing across cliffs streaked red and black from eons of falling water drawing out embedded minerals. Natural figures of people emerge in these streaks of color, figures even more gigantic than the human-made paintings in the caves and similarly elongated and boxy, colored half red, half black. We observe these giants in motion under the projector light of the sun and wonder how long the painters stared at these walls before they began to recreate the animations in their own fashion on the canvas of the landscape.

Juan Carlos looses Rocío and the burros to roam free, though he stakes his mule and Enriqueta's on leads long enough to forage. Throughout the afternoon we hear the chiming of the jenny's bell up the arroyo, first in one direction, then another. But no, Juan Carlos shakes his head. That is not her bell but the bell of a burro from a ranch to the east. Or to the west. Or his neighbor's burro. All are left to range freely until they're needed for work. To Enriqueta and me, the chimes sound identical, but to Juan Carlos they are as distinctive as the voices of his children.

When we ask why the other two mules are tethered, he explains that both have a horse mother and a donkey father, and they therefore believe themselves to be horses and want nothing more than to return to the ranch to be with the other horses. Rocío, though, is the son of a donkey mother and a horse father and believes himself to be a donkey, and he is happiest roaming free with the other donkeys, all of whom consider the desert home.

Toward dusk, Juan Carlos tracks the ringing of his jenny's bell up the arroyo, past the two caves, and far off the trail. His burros have no desire to be found, and we hear the chimes fading from us about as fast as we approach. But this is a vaquero's job, after all, and Juan Carlos tracks patiently. He shows us Rocío's footprints and the lead jenny's footprints. He shows us the footprints of burros from other ranches. Eventually he tracks his animals to a hiding place behind a multitrunked garambullo cactus, where, silent and wary, mouths full of saltbush, they stare back at us. In just a few short hours, we have become strangers to them.

With an elegant toss of a rope, Juan Carlos lassos the jenny and leads her downhill. The others follow. The route is familiar to them, since all trails lead to the *tinaja*—the only permanent water hole in the arroyo, a natural well bored by erosion into the bedrock of the dry riverbed in a circular vertical shaft the size of an elevator. Its edges are made slippery on all sides by the wear of many feet of many species over many centuries. At present the water is too low and the well too steep for the animals to drink without danger of falling in. So we lower our plastic food crates on ropes and haul water back to the surface.

In near darkness, we watch as Rocío and the burros visibly regenerate the wild limbs of their natures, dancing on nervous feet, snorting their fears. We coax them to drink, though they will do so only at the outmost stretch of their necks, with the whites of their eyes showing and their ears laid back. As quickly as camels, they inhale water through the bellows of their muzzles, pumping gallons in seconds.

And then they're gone, stealing into darkness on their slipperlike unshod feet. We are tempted to follow, drawn by the seductions of the wild donkey mother.

It's a hard climb to the Cueva de los Corralitos on a crumbling vertiginous trail requiring hands and feet and some unintentional ground-level study of cactuses. At one point, picking myself up, I come nose to nose with a miniature cave no bigger than a tent, tucked under a rock shelf and brilliantly animated with paintings of hares and humans. Juan Carlos has never seen it before.

We climb nearly to the top of the canyon, to a two-hundred-foot-long rockshelter with arched roof and benchlike floor opening to panoramic views of distant peaks and lush desert groves. The interior is dark and deep enough to hold a dank chill even in the heat of midday. It's clear the people did more than paint here, though the sloping ceiling overflows with enormous figures. The innermost reach of the cavern, where its ceiling meets the floor, is blackened by centuries of soot from cooking fires, and throughout the cave lie the little stone circles, *corralitos,* for which the site is named, constructed long ago for purposes now forgotten. The floor of the cave is littered with the dung of burros and mules who wander up here in pursuit of shade and leave behind the scent of barn.

The murals overhead are enormous, fragmentary, multilayered, jumbled. The painters painted on top of other paintings, so the screen of the cave appears to be showing many films at the same time. The oldest images have been overlaid nearly beyond recognition. Eagles rise on rock skies, on cruciform capes of spiky wing and tail feathers. A headless red mountain lion prowls among deer impaled on spears or arrows. Tuna and whales swim through volcanic waves, past a snake and a moray eel. Armies of half-red, half-black silhouetted humans face us, forearms raised to the sky, fingers spread, some wearing feathery starbursts on their heads. The human females are denoted by breasts emerging from their armpits. In the center of the rockshelter stands a four-foot-tall stone pedestal painted red, its top surface as shiny as polished coral.

A few years from now, visiting a strikingly similar rockshelter at the confluence of the Pecos and Rio Grande rivers in southern

Texas, I see a similar pedestal likewise polished to a high luster and surrounded by enormous paintings, including boxy, elongated, red-and-black human figures with arms raised. My visit to the Fate Bell Shelter in Texas's Seminole Canyon State Park is facilitated by a weather-beaten docent with a penchant for storytelling. The pedestal in the Texas cave was used for the sacrificial butchering of game, he says, and was polished smooth by centuries of bloodletting. Chewing a toothpick, squinting from under a cowboy hat, eyes twinkling, he tackles the meaning of the paintings, weaving epic tales of shamans and young men, of peyote-infused hallucinations, of trips into the minds of deer and mountain lions, into the realm of death and back again. His stories are far more vivid than the paintings themselves, since Fate Bell's four-thousand-year-old images have been severely eroded by the suddenly humid climate brought on by the damming of the Rio Grande.

The winter sun is setting as he leads the way out of Seminole Canyon. I tarry at the precipice, letting his voice and his stories fade away. The world I came to see stretches below—the canyon bored as smooth as a bowling alley by flash floods, the bedrock stamped with the fossilized footprints of dinosaurs. The rockshelter stares back, a Cyclops in limestone. My own shadow casts a long and boxy stencil on the far wall. Without thinking, I raise my arms, hands open.

Juan Carlos is a quiet man, and he leads us in silence through an arroyo in the Sierra de San Francisco to a slit cave painted with striking human figures and deer segmented by stripes and scales. Countless petroglyphs are drilled and carved into the soft tuff below the ceiling of the cave: abstract circles and carved parallel lines as dense and repetitive as newsprint.

The highlight of this cave is one of the most modest paintings on view anywhere in Baja California: a small depiction in ochre of a childlike sun, with lines radiating from a circle, nestled beside the outline of another circle more than half filled with ochre pigment. It would be easy to pass this by as something inconsequential in comparison to imposing murals of seven-foot-tall humans and thirteen-foot-tall deer. But Juan Carlos whispers to us to look again, to see what many did not see for many centuries: the depiction of a supernova.

The story of this image has a long lineage, and the starting place for its rediscovered meaning dates back to the year 1054, when Chinese astronomers noted a guest star in the constellation Taurus and recorded that its glow was visible in the daytime sky for twenty-three days and in the nighttime sky for six hundred fifty-three days. Similar observations were noted by Japanese, Persian, and Arab astronomers.

Little more thought was given to this celestial light for a long time. It was not noted in 1731 when the English doctor and astronomer John Bevis first observed a nebulous cloud within our own Milky Way galaxy nor, more than a century later, when another English astronomer named it the Crab Nebula. The visit of the guest star was nearly forgotten until the early twentieth century, when—working backward in time to calculate the rate of expansion—astronomers surmised that the Crab Nebula was the remains of the 1054 supernova observed by Chinese and Arab astronomers.

Later the American astronomer William Miller calculated that the 1054 supernova appeared in western North America in dazzling conjunction with a crescent moon. He correlated this sight to two pieces of prehistoric rock art in Arizona, each depicting a star beside a crescent moon. Later astronomers found strikingly similar rock art of conjunct stars and crescents at other sites in the American Southwest. In 1971 the explorer Harry Crosby, traveling by mule in the Sierra de San Francisco, came upon this image of a star and a moon—the only painting of its kind in the murals of Baja California, which he later surmised was also an image of the 1054 supernova.[1]

We sit quietly, sheltering from the sun under the overhang of the cave, beneath the ochre star and ochre moon. Enriqueta leans against the back wall, staring beyond the paintings in the foreground to the desert in the background. Juan Carlos lies on his back on the floor of the cave, using my binoculars to bring distant paintings into closer view. I sit on the edge of the cave, looking in at my fellow travelers and at the images made by forgotten people on rock welled from the Earth millions of years ago.

The 1054 supernova occurred 6,300 years before anyone on Earth witnessed it. The explosion dismantled a star more than 37,000 trillion miles away from us. The blast radiated as much energy as our sun will emit in the course of its life, and its light traveled at the fastest speed possible, the speed of light itself, yet it still took more than sixty centuries to get here.

In the Sierra de San Francisco, the cataclysm is reported with a childlike star beside a crescent moon. Yet this humble image cinches the threads in a long necklace of human understanding, of Chinese writings, Persian observations, Arabic records, English telescopes, American mathematics. It reaches through time to link distant space to the space here, connecting the people then to the people now.

In the late eighteenth century, landscape painters made use of an optical device known as a black mirror—a reflecting glass with a convex face slightly darkened. The idea was for the painter, amateur or professional, to turn her back to the landscape, suspend the black mirror beside her canvas, and paint the reflected image of the landscape. The black mirror encompassed a wider field of view and a simplified perspective, grading tones, downplaying contrasts, washing colors into a muted, luminous harmonic.

In the Cueva de la Supernova, we turn to face the rock and the image from a dark mirror of the human past. On convex walls darkened by soot we read the broader perspective—of disparate, exploded realms of sea, air, land, reassembling into wholes. We see the numinous interplay among whales, pronghorn, sardines, lynx, eagles, sea turtles. We see ourselves and know that we are part of the landscape, the seascape, the airscape, and can never be apart from them.

When we turn away from the reflected and back to the real, the silent stories playing in the theater of our minds encourage our human wisdom to be wild again, and wiser.

A cold finger trickles into the rockshelter. While we have been lost in the past, a shift has occurred outside. Windblown blue skies have ceded to gray altostratus mottled with hidden sunlight. A low-pressure system is dropping in from the Pacific. It is likely raising dust storms from the vast salt pan of the Laguna Salada, the Great Black Dry Lake, which the Seri called Caailipoláacoj. It may well be blowing stragglers to the shores of Isla Rasa, a hundred miles north of us. Its perturbations are rippling the rug of the atmosphere all the way to Newfoundland—and beyond, all the way below, tugging at the waters, redirecting their currents and the currents below them, raising the depths to the surface.

On the walk back to camp we see the northern fringe of clouds shedding their moisture in streaks of virga—precipitation that never reaches the ground. These streaks are infused with rain-making nuclei from the emissions of phytoplankton. They are likely populated with ice-nucleating bacteria from soil, snow, and seedlings, ascending perpetually on the conveyer belt of updrafts to pollinate the atmosphere and make the rain that scatters the bacteria to the far-flung reaches of the world, to colonize weather with life, and life with water, and water with weather, before beginning again.

The deep blue home arises and falls in every molecule of our planet. The tides within us never cease.

Acknowledgments

My oceangoing ancestors plied the seaways of the world and bequeathed to me some extra measure of restlessness that launches me offshore whenever life threatens to dry out. I'm indebted to them all, known and unknown, including my far-traveling parents, Mark and Patience Whitty, who taught me to love the sea by taking me there every summer and giving me the freedom to run wild along its shore. It's a gift I wish all children could receive. I'm also grateful to have shared a wild childhood shore with my wild brothers Tim and John.

Many thanks to my wise editor at Houghton Mifflin Harcourt, Deanne Urmy, for her insightful touch throughout this book and for her generous enthusiasm that helped keep a writer afloat in stormy weather. Ongoing thanks to Clara Jeffery at *Mother Jones* for tossing me a line and enabling me to step between a vessel called *Filmmaking* and one called *Writing*. I'm grateful to Melanie Jackson, my agent, for tolerating both my long silences and my squalls of activity. Thanks as well to Peg Anderson, my copyeditor, who freed me from the traps of circulating eddies and quieted rough writing. I appreciate the ongoing assistance of associate editor Nicole Angeloro at Houghton Mifflin Harcourt.

I am further indebted to all those I've sailed with over the years, more than I can mention here, though specifically to those who do appear in these pages, including two I've known longest, Enriqueta Velarde and Hardy Jones. Both helped me learn to see the world the way I see it now.

On the research front, I believe I owe Mark Hixon a few hundred beers in return for the many articles he has kindly retrieved for me from the unaffordable wealth of scientific journals and for his friendship and contagious knowledge of the water world. I owe more than a few beers as well to Lisa Levin and Victoria Orphan for correcting my writing on the cold methane seeps and other extremophile communities of the deep and for allowing me the extraordinary opportunity of sailing with them aboard *Atlantis*. Thanks to them also for forgiving my flawed dubbing of the *Alvin* dive tapes, for which I still feel as lowly as a dorvilleid. My gratitude as well to Ruth Curry for allowing me to sail with her aboard *Atlantis* during that memorably miserable cruise.

My love and thanks to Ian Riedel, who endured the strange voyage required for me to write this book. I was mostly at sea for more than a year, though only a metaphorical sea, asail on a lonely currach of the mind, trying to find my way. Throughout, from the other side of my office door, he anchored me.

I have had the inestimable good fortune to spend my adult life working in wild places alongside the wild beings who make our world livable and who have endowed my tenure with uncommon joy, inspiration, wonder, laughter, ideas. I am profoundly grateful that our planet includes a nonhuman world and that my relationship to it helps me become a better human.

Notes

The Very Air Miraculous

1. Steinbeck's classic tale of travel, science, and comedy, *The Log from the Sea of Cortez* (Viking Press, 1951), was his second attempt to tell the story of his 1940 journey with Ed Ricketts. The first book, *Sea of Cortez: A Leisurely Journal of Travel and Research* (Viking Press, 1941), cowritten with Ricketts, drew heavily on Ricketts's journal entries as well as his species list of marine life. That book never sold well, and three years after Ricketts's death Steinbeck revised it, adding his indelible portrait of the marine biologist and subtracting the species list.

2. Arthur Cleveland Bent's *Life Histories of North American Gulls and Terns* (1921) was the second of the twenty-one volumes that consumed most of his adult life. Publication of the volumes spanned forty-nine years, with the last two installments published posthumously.

The River That Was Nowhere and Everywhere

1. A scientific overview of Rasa's breeding ecology can be found in the chapter "Breeding Dynamics of Heermann's Gulls," by Enriqueta Velarde and Exequiel Ezcurra, in *A New Biogeography of the Sea of Cortés,*

ed. Ted J. Case, Martin L. Cody, and Exequiel Ezcurra (Oxford University Press, 2002). This invaluable book compiles a wealth of scientific knowledge of the region's flora and fauna.

2. In *The Future of Life* (Alfred A. Knopf, 2002), Wilson employs a compelling mix of science, philosophy, and fiction—notably in the Prologue, where the author holds an imaginary conversation with Henry David Thoreau on the state of Walden Pond and Earth itself today.

3. For more on the loss of biodiversity, see my article "Gone: Mass Extinction and the Hazards of Earth's Vanishing Biodiversity," *Mother Jones,* May/June 2007.

4. The Red List is fully catalogued online and includes species accounts as well as links to bibliographies, data, and photos: http://iucnredlist .org/.

5. James R. Spotila, *Sea Turtles: A Complete Guide to Their Biology, Behavior, and Conservation* (Johns Hopkins University Press, 2004), summarizes the life history, ecology, and conservation status of the world's sea turtles.

6. Leopold's luminous description of the extinct delta of the Colorado River is included in his collection of essays, *A Sand County Almanac* (Oxford University Press, 1949).

Hunger Island

1. Mark H. Carr, T. W. Anderson, and M. A. Hixon, "Biodiversity, Population Regulation, and the Stability of Coral-Reef Fish Communities," *Proceedings of the National Academy of Sciences* 99, no. 17 (2002): 11241–45.

2. Thomas Bowen's book, *Unknown Island: Seri Indians, Europeans, and San Esteban Island in the Gulf of California* (University of New Mexico Press, 2000), details a fascinating history of guano mining throughout the Midriff Islands, including Isla Rasa.

3. Griffing Bancroft's long-out-of-print book *The Flight of the Least Petrel: Lower California: A Cruise* (G. P. Putnam's Sons, 1932) is an account of his ornithological cruise, long before Ricketts and Steinbeck's, including his visit to Isla Rasa. Bancroft took a peculiar dislike to the island's breeding terns: "Rattled and brainless, their minds ap-

pear weak to the point of representing the minimum of avian mentality. They cause me to wonder why evolution has been so inconsistent as to have created such beautiful things and then to have endowed them with so little intelligence."

4. A treasure trove of scientific information is contained in the online archive the Birds of North America Online, Cornell Lab of Ornithology and American Ornithologists' Union.

One Hundred Days of Solitude

1. Translation from *The Sea of Cortez: A Place with a Future* (Pulsar, 1998), by Alejandro Robles, Exequiel Ezcurra, and Cuahtémoc Léon.

Whorls

1. Gary Paul Nabhan's account of Seri life, *Singing the Turtles to Sea: The Comcáac (Seri) Art and Science of Reptiles* (University of California Press, 2003), includes many examples of the Seri's exhaustive nature knowledge, much of it embedded in their unique language, which is an isolate. Nabhan provides a vocabulary.

Mirage

1. Brown's observations are included in Arthur Cleveland Bent's *Life Histories of North American Petrels and Pelicans and Their Allies* (1922; reprint, Dover, 1964).

Emotional Ecology

1. Anthony's observations appear in Arthur Cleveland Bent's *Life Histories of North American Petrels and Pelicans and Their Allies*.

The Distant Geography of Water

1. See Peter L. Lutz, J. A. Musick, and J. Wyneken, eds., *The Biology of Sea Turtles,* vol. 2 (CRC Press, 2003), which includes chapters on ancient and historic interactions between sea turtles and people, turtle morphology, sensory biology, reproduction, physiology, migrations, feed-

ing ecology, life histories, population ecology, conservation, and sea turtle management.

The Ecumenical Sea

1. I drew heavily on John Lindow's fascinating and scholarly accounts and interpretations in his *Norse Mythology: A Guide to the Gods, Heroes, Rituals, and Beliefs* (Oxford University Press, 2001), for my depictions of Ægir, the other *jötnar,* and the Scandinavian gods. His chapter "The Nature of Mythic Time" provides a haunting window into the mind of the Norse past.

The Tempest from the Eagle's Wings

1. This version of Sæmund's verse appears in *Legends and Superstitions of the Sea and of Sailors in All Lands and at All Times* (Belford, Clarke and Co., 1885), by Fletcher S. Bassett, Lieutenant, U.S. Navy.
2. Farley Mowat's *Sea of Slaughter* (McClelland and Stewart, 1984) is a searing account of the brutal decline of biodiversity along Canadian shores. Mowat signed my copy "Up the Animals!!"—a memento of an unforgettable day filming him in his Nova Scotia home, after which he genially slammed a whiskey bottle onto his kitchen table and roared, "I worked for you all day. Now you'll drink with me."

One Meritorious Act

1. Peterson and Fischer's 1955 book, *Wild America* (reprint, Mariner Books, 1997), includes an engaging account and charming sketches of their long hike in 1953 across the boggy landscape to and from Cape Saint Mary's. Thirty-two years later I had the great good fortune to spend a day filming Roger Tory Peterson at his home in Old Lyme, Connecticut. I was the junior member of the film crew and shyly confessed toward the end of the day that I was approaching four hundred identified species on my birder's life list. Peterson was genuinely thrilled to hear of my modest accomplishment and twice repeated it to his wife, Virginia.
2. Carson's enormously influential *Silent Spring* (Houghton Mifflin, 1962)

reinvigorated a slumbering environmental movement and helped it grow into a global force in the latter half of the twentieth century. That force, still building today as it merges with science, is likely to be known someday as the Environmental Revolution, with Carson among its most effective liaisons between science and the public.

3. Townsend's account appears in Arthur Cleveland Bent's *Life Histories of North American Petrels and Pelicans and Their Allies,* p. 219.

Jump Cut

1. "Big Old Fat Fecund Female Fish: The BOFFFF Hypothesis and What It Means for MPAs and Fisheries Management," *MPA News* 9, no. 3 (2007): 1–2.
2. Arthur Cleveland Bent, *Life Histories of North American Diving Birds* (reprint, Dover, 1964), pp. 93–94.

Lament for the Thirty Million

1. Francis K. Wiese, G. J. Robertson, and A. J. Gaston, "Impacts of Chronic Marine Oil Pollution and the Murre Hunt in Newfoundland on Thick-Billed Murre *Uria lomvia* Populations in the Eastern Canadian Arctic," *Biological Conservation* 116 (2004): 205–16.
2. Sabina I. Wilhelm, G. J. Robertson, P. C. Ryan, and D. C. Schneider, "An Assessment of the Number of Seabirds at Risk During the November 2004 *Terra Nova* FPSO Oil Spill on the Grand Banks," *Canadian Wildlife Service Technical Report Series,* no. 461, September 2006.
3. R. A. Khan and P. Ryan, "Long Term Effects of Crude Oil on Common Murres (*Uria aalge*) Following Rehabilitation," *Bulletin of Environmental Contamination and Toxicology* 46 (1991): 216–22.
4. Francis K. Wiese and Pierre C. Ryan, "Trends of Chronic Oil Pollution in Southeast Newfoundland Assessed Through Beached-Bird Surveys 1984–1997," *Bird Trends Newsletter,* no. 7 (1999), Canadian Wildlife Service.
5. R. G. Butler, A. Harfenist, F. A. Leighton, and D. B. Peakall, "Impact of Sublethal Oil and Emulsion Exposure on the Reproductive Success of Leach's Storm-Petrels: Short and Long-Term Effects," *Journal of Applied Ecology* 25 (1988): 125–43.

All Time Is Now

1. Peter Beamish's Web page describing his research is at http://ocean contact.com/research/WhatisRBC.html.

Trophic Cascade

1. R. A. Myers and B. Worm, "Extinction, Survival or Recovery of Large Predatory Fishes," *Philosophical Transactions of the Royal Society B* 360 (2005): 1453.

2. C. T. Darimont et al., "Human Predators Outpace Other Agents of Trait Change in the Wild," *Proceedings of the National Academy of Sciences* 106, no. 3 (2009): 952–54.

3. Kenneth T. Frank, B. Petrie, J. S. Choi, and W. C. Leggett, "Trophic Cascades in a Formerly Cod-Dominated Ecosystem," *Science* 308 (2005): 1621–23.

Bone Rafters

1. Robert McGhee's *The Last Imaginary Place: A Human History of the Arctic World* (Oxford University Press, 2005) provides a fascinating and intimately detailed picture of Norse life in the New World. I've drawn heavily upon his knowledge in these pages, including my description of the Dorset culture and the Inuit expansion east during the Medieval Warm Period.

2. Nancy Marie Brown tells Gudrid's story and much more in her wonderfully wide-ranging book *The Far Traveler: Voyages of a Viking Woman* (Harcourt, 2007).

3. From *Beyond Remembering: The Collected Poems of Al Purdy* (Harbour Publishing, 2000). It is one of several haunting poems Purdy wrote about the people of the far north.

Soundsabers

1. From Whitehead and Weilgart, "The Sperm Whale: Social Females and Roving Males," one chapter in an extraordinary compilation of scientific knowledge about whales and dolphins edited by Janet Mann et al., *Cetacean Societies: Field Studies of Whales and Dolphins* (University of Chicago Press, 2000).

2. Pitman and Chivers's scientific description of this harrowing account is "Killer Whale Predation on Sperm Whales: Observations and Implications," *Marine Mammal Science* 17, no. 3 (2001): 494–507.

3. R. L. Pitman and S. J. Chivers, "Terror in Black and White," *Natural History*, December 1998.

The Existence of a World Previous to Ours

1. Joe Roman et al., "Whales Before Whaling," *Science* 301 (2003): 508.

2. Jeremy Jackson's chapter "When Ecological Pyramids Were Upside Down," in *Whales, Whaling and Ocean Ecosystems,* edited by James Estes et al. (University of California Press, 2007), is one of many poignant scientific examinations of the world at the time when whales were numerous. The information I present regarding killer whales and the decline in Steller sea lions and sea otters, as well as the decline of North Atlantic right whales, also comes from this book.

3. Daniel F. Doak, T. M. Williams, and J. Estes, "Great Whales as Prey," in *Whales, Whaling, and Ocean Ecosystems,* pp. 242–43.

4. Craig R. Smith and A. R. Baco, "Ecology of Whale Falls at the Deep-Sea Floor," in R. N. Gibson and R. J. A. Atkinson, eds., *Oceanography and Marine Biology: An Annual Review* 41 (2003): 311–54.

Nemesis

1. For more on this voyage, see my article "The Fate of the Ocean," *Mother Jones,* March/April 2006, which includes many other issues affecting the World Ocean, from dead zones to overfishing to climate change.

The Inexplicable Waves

1. For more information on this, see my article "The Thirteenth Tipping Point," *Mother Jones,* November/December 2006, which examines the interactions between the twelve planetary tipping points and the single antidote: the thirteenth tipping point, ourselves.

2. William K. Dewar, "Oceanography: A Fishy Mix," *Nature* 460 (2009): 581–82.

At the End of Hunger

1. This translation can be found in *Ovid's Metamorphoses in Fifteen Books, Translated by the Most Eminent Hands*. The original 1717 edition, printed for Jacob Jonson, was reissued by the Limited Editions Club in 1958 under the title *Ovid's Metamorphoses. In Fifteen Books. Translated into English Verse under the Direction of Sir Samuel Garth by John Dryden, Alexander Pope, Joseph Addison, William Congreve and Other Eminent Hands*. The entire text is available online at the Internet Classic Archive.

2. I found a transcription of Bishop John Inglis's interview with Shanawdithit online, published by Memorial University, in pages about the native religions of Newfoundland and Labrador: http://www.mun.ca/rels/native/beothuk/ingshan.html.

Black Mirror

1. Crosby's illustrated book, *The Cave Paintings of Baja California: Discovering the Great Murals of an Unknown People* (Sunbelt Publications, 1998), details his explorations over four decades and more than a thousand miles in the saddle in search of the lost cave art of Baja California.